Biker

Biker

Truth and Myth:

How the Original Cowboy of the Road Became the Easy Rider of the Silver Screen

Bill Osgerby

PROSPERO BOOKS

This edition produced in 2005 for Prospero Books

ISBN 1-55267-742-7

Printed and bound in China.

10 9 8 7 6 5 4 3 2 1

This book was conceived,
designed, and produced by
THE IVY PRESS LIMITED
The Old Candlemakers
Lewes, East Sussex BN7 2NZ, UK

Creative Director: *Peter Bridgewater*
Publisher: *Sophie Collins*
Editorial Director: *Jason Hook*
Art Director: *Karl Shanahan*
Senior Project Editor: *Caroline Earle*
Designer: *Simon Goggin*
Picture Research: *Shelley Noronha*
Photography: *APM Studios*

Contents

INTRODUCTION

The Last American Hero

Biker Mystique

Rebellious riders of the storm, the biker brotherhood represents a beguiling mix of myth and reality.

The biker stands tall as one of modern America's most powerful icons. The last American hero/antihero, the biker's image of trenchant independence and rugged individualism harks back to the mythic figures of the cowboy and the Western pioneer. Davie Crockett, Wyatt Earp, and Jesse James have long since bitten the dust, but the biker has assumed their mantle. Just as the hardy frontiersman personified the sense of personal freedom and robust self-reliance at the heart of the American Dream, the biker has come to symbolize American ideals of confidence, resourcefulness, and individual liberty.

Sure, he ain't always been pretty. Sometimes he's been out-and-out mean and ugly. But, despite his rough edges, the biker represents the new pioneer-hero. Defiant and unfettered by mainstream conventions, the asphalt cowboy has always gone his own way. Yet he's also been an independent and freethinking patriot—an enigmatic roughrider who has been steadfastly loyal to the nation's core values and heritage.

Tough, resourceful, and independent, the biker is the modern-day heir of the frontier pioneer.

The Biker—Myth, Legend, and Reality

Myth, legend, and reality overlap in the history of the biker. To a large extent, the biker is a creation of popular culture—a glamorized product of movies, magazines, and pulp fiction. But he is not pure invention. Myths always form around a kernel of truth, and popular representations of the biker are no exception. Dissident outsiders, bikers have always resisted the humdrum regime of work and home, and instead have struck out on the highway to adventure, freedom, and camaraderie.

Romance and myth have also played important parts in the biker story. Pulp fiction hacks, tabloid newshounds, and Hollywood hucksters have all exploited the biker's compelling persona, giving an added kick of melodrama to the image of the motorcycle renegade. This media spin has, in turn, influenced the biker's self-perception and outlook, while the media's fascination with motorcycle subculture has helped popularize the style and attitude of the American biker worldwide.

A Full-Throttle Saga

Biker takes you on a wild ride through the history of the biker myth and its impact on modern culture. Charting the development of the bad-to-the-bone biker as a preeminent American icon, this book shows how the biker's aura of reckless rebellion has been a charismatic force that has resonated through popular culture the world over. We'll see how popular media have always been fascinated with motorcycle subculture, offering representations of the biker way of life that are a high-octane mix of fact and fantasy, gritty reality, and alluring romance.

Biker chronicles the development of the biker brotherhood from its beginnings, among loose fraternities of boisterous riders in the aftermath of World War Two, through to the global proliferation of sophisticated biker gangs during the 1980s, 1990s, and beyond. We'll see how this history has been reflected in changing representations of the biker across a spectrum of literary fiction, movies, magazines, and a host of other media. The book highlights the way Hollywood glamor and pulp exploitation alike have been inspired by the thrilling exploits of maverick motorcyclists, and it shows how the popular imagery has, in turn, fed back into the development of the biker mystique. *Biker* is the full history of this interaction, showing the inextricable links between the lifestyle and lore of the biker and his representation in popular culture.

Johnny

01

REVVING UP

The Origins of Biker Mystique

Harley-Davidson: Birth of an American Icon

Ask anyone; they'll all say the same. Nothing else has the growl of a Harley-Davidson "V-twin" motorcycle engine. Its deep, resonating rumble is something you feel as much as hear. No other bike can quite match the way the Harley's roar evokes the white-knuckle freedom of the open road and, for over a century, bikers across the world have venerated the history and traditions of the Harley-Davidson Motor Company—or, as it's known to insiders, simply "The Motor Company."

The Harley's rough-and-ready beginnings are rooted in Milwaukee, Wisconsin. In 1903, 23-year-old William Harley and 20-year-old Arthur Davidson began tinkering with ideas to take the slog out of bicycling. Soon joined by Arthur's brothers, Walter and William, the team worked in a 10 x 15-foot (3 x 4.5-meter) wooden shack (with the legend "Harley-Davidson Motor Company" crudely scrawled on the door) to develop their first motorcycle. With a top speed of around 35 mph (55 km/h), the first production Harley-Davidson was little more than a 410cc single-cylinder engine mounted on a bicycle frame. But business soon picked up. By 1906 the company had a full-time workforce of six, and a new factory was set up on the Juneau Avenue site that would become Harley-Davidson's established home. By the end of the year no less than 50 bikes had rolled off the Harley production line.

The Big "V" Unleashed

With improving sales, the Harley-Davidson Motor Company was incorporated on September 17th, 1907. The same year saw the first Harley-Davidsons sold for police duty, and the firm's workforce and factory doubled in size. As production grew, new models took shape. In 1909 Harley-Davidson's distinctive twin cylinder engine was unveiled—the image of two cylinders in a 45-degree "V" configuration was to become one of the most enduring biker icons. The performance of the original 1909 "5D" model was disappointing but, within two years, the more powerful "7D" version had been launched and the big V-twin was here to stay. In 1910 the famed Harley-Davidson "Bar and Shield" logo also made its first appearance. Trademarked at the US Patent Office the following year, it was destined to become a sacred totem of American motorcycling.

THE "SILENT GRAY FELLOW"

Priced at $200, the "Silent Gray Fellow" was the first Harley-Davidson motorcycle to be produced in significant numbers. Launched in 1906, the single-cylinder machine got its name from its signature gray paint scheme and quiet running at a time when loud pipes were considered a drawback. "Fellow" denoted the reliability of the bike, promoted as a good and faithful friend to its owner. By 1908 Harley-Davidson's bikes had become stouter, with improved front forks and wider fenders to accommodate wider tires. Production of single-cylinder Harleys ended in 1918, as the company switched its attention to bigger V-twins.

Since 1903 Harley-Davidson motorcycles have embodied the freedom of the open road.

THE CREATION OF HARLEY-DAVIDSON

The legendary American Harley-Davidson motorcycle company was founded in 1903 in Milwaukee, Wisconsin by William Harley and the Davidson brothers—Arthur, Walter, and William. William Harley and Arthur Davidson had previously worked full-time for a local engineering firm, but tinkered with engines in their spare time. Their initial aim was to produce a marine outboard motor, which would allow them to reach their favorite fishing spots more easily. But they were soon captivated by a new invention—the motorcycle.

Arthur's brother Walter was working as a railroad machinist in Kansas but soon joined the original partners. The team built only three bikes in the first year, but by 1905 production had crept up to eight. The Davidsons' aunt, Janet Davidson, was responsible for the decorative features on the first bikes, creating the Harley-Davidson logo for the gas tank and applying the striped trim. For the first three years Harleys were painted black, with gold striping and red lettering.

In 1907 the third Davidson brother, William, joined the firm. On September 17th the Harley-Davidson Motor Company was incorporated, with the company's stock split between the four founders. Walter Davidson was appointed company president, Arthur Davidson the sales manager, William Davidson the factory manager, and William Harley the chief engineer.

The Motorcycle Pioneers

William Harley and the Davidson brothers are star-spangled legends, but they weren't the first motorcycle pioneers. During the late 1860s one Sylvester Howard Roper of Roxbury, Massachusetts, had already demonstrated his two-wheeled, steam-powered "velocipede" at New England fairs and circuses. But it is Gottlieb Daimler, a German car-builder, who is in fact credited with building the first modern motorcycle, in 1885. Daimler (who later teamed up with Karl Benz to form the Daimler-Benz Corporation) developed a small, single-cylinder gasoline engine suitable for universal application, and he quickly had the idea of fitting it to a two-wheeled wooden frame.

The German engineer Gottlieb Daimler (1834–1900), pioneer of the internal combustion engine and the automobile, and grandfather to the modern motorcycle.

The wheels of Daimler's bike were iron-banded and wooden-spoked and—for stability—smaller, spring-loaded outrigger wheels were attached on either side. Daimler's young son, Paul, was the first to take the contraption on a test drive. Daimler's daughter also took the vehicle for a spin, but was less adept than her brother and plowed the world's first motorcycle into a nearby tree.

The first successful production bike was also German. Developed by Heinrich and Wilhelm Hilderbrand and Alois Wolfmüller in 1894, it was powered by a large, water-cooled, four-stroke gasoline engine mounted low on the frame, and boasted the latest pneumatic tires, recently invented by John Boyd Dunlop. At around the same time a pair of French pioneers, Comte Albert de Dion and his partner Georges Bouton, developed an engine that made the mass production of motorcycles truly possible. Mounted on a tricycle, the Frenchmen's single-cylinder, four-stroke engine had a capacity of around 125cc. Dion and Bouton successfully licensed production of their engine in England, Germany, Belgium, and the United States, but it was also copied by numerous firms as motorcycle manufacture spread worldwide.

The Motorcycle Takes Shape

In the pioneering days many different motorcycle layouts existed. Some mounted the engine either above or to the side of the front wheel; others mounted it above or behind the rear wheel. The problem of connecting one of the wheels to a continuous belt drive proved a major stumbling block. An effective solution was devised in 1901 by two talented Russian exiles living in Paris—Michel and Eugene Werner. Their patented "New Werner" layout placed a 230cc engine low in a diamond-shaped, steel frame midway between the two road-wheels, while a leather belt drive connected the engine to the rear wheel. Improving rider stability through its lower center of gravity, the configuration proved a big success and provided the blueprint for motorcycle design for the next 100 years.

The Indians are Coming

Many of the early advances in motorcycle design were made in Europe, but American innovators were also important. Oscar Hedstrom and George Hendee, for example, were leading US pioneers. Meeting through their mutual interest in motorcycle racing,

Oscar Godfrey, winner of the 1911 Isle of Man Senior TT, posing with his Indian motorcycle at the British Motor Cycle Race Club trials at Brooklands.

Hedstrom and Hendee founded the Indian Motorcycle Company of Springfield, Massachusetts, in 1901. The name "Indian" was chosen to underline the firm's American pedigree, with Hendee taking charge of promotion and marketing as the "Big Chief" while Hedstrom handled engineering as the "Medicine Man."

The first Indian bikes utilized a bicycle-style, diamond frame, but by 1909 they were using the "loop" frame that became a distinctive feature of modern machines. Oscar Hedstrom also devised a more effective gearing system. After much experimentation, a two-speed, center-shaft transmission was perfected. Incorporating a clutch, and integrated with a drive chain and rear-wheel gear sprockets, it was a winning formula. Stealing a technological lead over their rivals, Indian achieved remarkable success in the 1911 Isle of Man Tourist Trophy (or TT) races, which had been established four years earlier, scooping first, second, and third places.

Subsequent years saw Indian go from strength to strength. In 1913 it was producing nearly 32,000 bikes a year, and by 1914 Indian accounted for a 42 percent share of the American motorcycle market. The firm also boasted the largest motorcycle factory in the world—covering a million square feet (90,000 square meters), the factory was affectionately known as the "Wigwam."

The All-America Hog

Alongside Indian and Harley-Davidson, other American motorcycle firms also sprang up. The Chicago-based firm Excelsior produced its first bike in 1907, and the Pierce Arrow Motor Company followed suit in 1909. In 1911 Henderson motorcycles was launched in Detroit by brothers Will and Tom—Will going on to found Ace Motorcycles in Philadelphia in 1919.

However, Harley-Davidson was the brand destined to become an integral ingredient of biker lore and lifestyle. Although rarely the best performing or most technologically advanced bikes, the Harley-Davidson's unique aura of American individualism and robust independence has found deep resonance in biker culture. Hefty, uncompromising, and brutish, the archetypal "hog" (as Harleys are fondly known) is a "frontier classic"—not simply a means of transport, but a talisman of adventure, freedom, and rugged self-reliance.

At the turn of the century two riders pose with their prized Harley-Davidson bike and sidecar outside Guy Webb's Minneapolis motorcycle shop.

The Motorcycle Goes to War

By 1914 Harley-Davidson's solid reputation was established. Harleys were not only the bikes chiefly deployed by American police forces, but nine different federal government departments also used Harley-Davidson machines—the Postal Service alone ran over 4,800 Harleys to help with its deliveries.

Astride his BSA motorcycle, a British dispatch rider prepares for frontline duty in 1916.

The early twentieth century saw the focus for motorcycle innovation shift from Europe to the United States. Motorcycles were no longer simply bicycles with engines, and design milestones such as the twist-grip throttle control, the drum brake on the rear wheel, the foot-operated clutch, and the starter motor were all perfected by American manufacturers.

All (Not So) Quiet on the Western Front

The outbreak of World War One in 1914, however, dampened the impetus for motorcycle development, particularly in Europe. The armed forces on both sides demanded reliability and durability from their machines rather than innovation. The civilian market also nose-dived, and many smaller companies were forced to close. But military demand was a boon to larger manufacturers, as thousands of bikes were drafted into service. European armies deployed motorcycles in a variety of roles, including reconnaissance and messenger vehicles and even ambulances. Others were equipped with special sidecars, which made a handy mount for the latest in machinegun hardware.

The US Army already had experience of the motorcycle's military potential. In 1916 General John J. Pershing's forces had used Indians, Excelsiors, and Harleys in their border skirmishes with the Mexican revolutionary, "Pancho" Villa. When the United States was drawn into World War One in 1917, the US Army's demand for motorcycles became voracious. Indian quickly slipped in to corner the military market, slashing their prices to secure wartime contracts and supplying 41,000 machines to Uncle Sam.

As Indian concentrated on supplying the military, Harley-Davidson cannily extended its grip on the civilian market. Reserving half its production for civilian customers, Harley's strategy afforded the company a huge commercial advantage once the war was over. Nevertheless, many Harleys did distinguish themselves in the field. The company produced over 15,000 army models for use in combat, communications, and transportation. And Harley-Davidson eagerly capitalized on the fact that one of the first Americans to enter Germany after the ceasefire—dispatch rider Corporal Roy Holtz—rode triumphantly into town astride his dependable Harley.

Locked 'n' loaded—members of the National Guard inspect the latest in motorcycle/machinegun hardware just before the breakout of World War One.

Nice bike, amigo. Mexican revolutionary leader Francisco "Pancho" Villa poses with an Indian motorcycle captured during skirmishes with US forces.

Roaring in the Twenties

By 1920 Harley-Davidson was the largest motorcycle manufacturer in the world. The company's Juneau Avenue factory—re-tooled and expanded to house a workforce of 2,500—was capable of producing 35,000 bikes a year. Harley also enjoyed major race success. Initially, the firm ignored national competition, but in 1914 Harley-Davidson established a racing team (soon dubbed the "Wrecking Crew") and launched a serious assault on national races. The team's first rider, Leslie "Red" Parkhurst, made a particular mark, winning prestigious events such as the 1916 Federation of American Motorcyclists national championship.

Whoa! An early Harley-Davidson is put through its paces in 1912.

Alongside other manufacturers, Harley-Davidson withdrew from factory-sponsored racing during wartime. But in 1919 Harleys returned to the track. That year "Red" Parkhurst earned his biggest victory in the 200-mile (320-kilometer) road race held in Marion, Indiana, while 1920 saw him break a fistful of speed records.

The Indian-Harley Wars

The 1920s, however, was a bleak decade for the American motorcycle business. Demand for bikes ebbed as automobiles became less expensive. With motorcycles increasingly marginalized from the mass market, bike manufacturers struggled to compete. And, as the competition intensified, Indian and Harley-Davidson became fierce rivals.

At Indian, despite Oscar Hedstrom's retirement in 1913, followed by George Hendee's in 1915, the Wigwam continued to produce excellent bikes. A new, side-valve engine known as the Powerplus entered production in 1916 and proved to be a reliable race-winner. Improvements to the Powerplus came with the launch of the 600cc Scout in 1920. In 1922 the engine was enlarged to 1000cc to become the Chief and, in 1924, enlarged further to become the 1200cc Big Chief.

Not to be outdone, Harley-Davidson also launched new models. In 1921 Harley's first 1200cc V-twin engine was introduced. In 1929, meanwhile, Harley launched a 750cc V-twin. Known as the "Flathead" due to its telltale, flat cylinder heads, the engine propelled Harley-Davidson to race victories throughout the 1920s. More attention was also given to styling. In 1925 the Harley gas tank became more rounded, with the arrival of the now-familiar classic "teardrop" shape.

Harley-Davidson also retained its shrewd approach to business. By 1928 the company had gained effective control of the American Motorcycle Association (AMA)—an organization founded in 1924 to oversee race meetings and promote US motorcycling.

The Depression's Rocky Road

American motorcycle manufacturers were hard hit by the Depression. By around 1929 the ranks of US motorcycle firms had thinned from as many as 200 to just three—Indian, Harley-Davidson, and Schwinn (manufacturer of the Excelsior and Henderson brands). With Schwinn folding in 1931, only Indian and Harley remained.

Indian's fortunes looked set to improve in 1929 when brothers E. Paul and Francis du Pont (of the DuPont chemical empire) bought shares in the company. The du Ponts set about streamlining the firm, and in 1934 the launch of the Sport Scout gave Indian a racetrack weapon that could compete well against the Harleys. In 1937, meanwhile, a new Sport Scout model heralded a stunning approach to styling, with gracefully rounded fenders and gas tank. Brilliant color schemes also became an Indian trademark, and soon Indians were available in more than two dozen gleaming, two-tone color combinations.

Indian had the looks, but Harley-Davidson had the engineering. In 1936, Harley leapt ahead through the introduction of an overhead-valve, V-twin engine known as the "Knucklehead," due to its four lumpy bolt housings on the chrome valve covers. Steadily improved, and enlarged from 61 to 74 cubic inches in 1941, the Knucklehead revived Harley's sales.

Indian's first four-cylinder bike on display at London's Olympia Motorcycle Show in 1927.

Promoting the Motorcyclist's Image

Harley-Davidson and Indian both strove to foster brand loyalty. Throughout the 1920s both camps' newssheets—*The Enthusiast* for Harley and the *Indian News* for Indian—encouraged employees, dealers, and customers to identify with the firm, and to see motorcycling as a sociable pastime. The AMA and enterprising dealers also began organizing motorcycle clubs. Thoroughly clean-cut, the clubs promoted an image of motorcyclists as respectable citizens. Against this, however, another image emerged. During the Depression many men drifted from town to town on weather-beaten bikes. These hobos on two wheels were often seen as scruffy troublemakers. The image of the motorcyclist as a sinister outsider would become still more prevalent after World War Two.

A World on Wheels

While Indian and Harley-Davidson were dominating the American motorcycle market, European firms were developing their own, often impressive, machines. Although the growth of the European motorcycle industry was slower than that in the US, by the beginning of the twentieth century numerous European firms had sprung up.

Riders hitch a lift on a Triumph and sidecar during the 1930s.

Britain, for example, saw the launch of such illustrious marques as Excelsior (founded in 1896), Triumph (1897), Matchless (1899), Ariel (1902), Norton (also 1902), and JAP (J.A. Prestwich, founded in 1903). The Birmingham Small Arms Company (BSA) also diversified, supplementing its arms production with motorcycle manufacture with the launch of its first single-cylinder, belt-driven machine in 1910.

Europe Gets Into Gear

Manufacturers also proliferated on the mainland continent. In Italy Benelli began producing motorcycles in 1911, followed by Garelli (1913), and Moto Guzzi (1921). In France, the René-Gillet Company was founded in 1898, while the Swedish arms firm Husqvarna and the Austrian firm Puch both began producing motorcycles in 1903. In Belgium, meanwhile, the FN company was a leading motorcycle pioneer, producing in-line, four-cylinder, and shaft-driven machines as early as 1903.

During the early twentieth century Germany also boasted thriving motorcycle firms. Wanderer was founded in 1902 and became well known for its high-quality bikes, while Hercules started business in Nuremburg, southern Germany, a year later. It was during the 1920s and 1930s, however, that German motorcycle manufacture really took off. At the end of World War One, the 1919 Treaty of Versailles severely restricted German production of aircraft and military equipment, forcing many manufacturers to find a new niche. For example, Zündapp (the Zünder und Apparatebau Company) was established in 1917 as a manufacturer of grenade and artillery fuses, but in peacetime was obliged to switch to motorcycle production.

BMW Kickstarts

The Bavarian Motor Works (BMW) started out as an aircraft company in 1917 (to this day, BMW's logo is a stylized white propeller against a blue sky). With the end of the war, the firm moved into the motorcycle (and automobile) business. Initially, BMW produced engines for other companies' machines, but in 1923 the company launched its own bike—the R3—featuring the now-famous horizontally-opposed "flat-twin" cylinders that became a distinctive feature of BMW motorcycles.

The interwar years were also a buoyant time for British bike production. At the first Olympia motorcycle show, in 1919, over 100 motorcycle firms exhibited their wares. Many were only assemblers: buying up parts and engines and building them into their own machines. But the number of companies grew steadily and in 1929—the peak year for British motorcycle production—some 147,000 machines were produced.

A Japanese Harley-Davidson?

Japanese motorcycle production, in contrast, initially developed slowly. By the 1920s Japan was a rapidly developing industrial power, but its output of motorcycles was virtually nonexistent. It was a situation Harley-Davidson turned to advantage. Facing a slump in US sales, Harley shifted its attention eastward, setting up an import operation in Japan that by the mid-1920s was selling more than 2,000 machines a year.

Using Harleys as all-purpose workhorses, the Japanese adapted the bikes for hauling all manner of light freight. The Japanese connection was developed further in 1935 when Harley-Davidson contracted the Sankyo Trading Company to manufacture complete machines under license. The result was the Rikuo—or "King of the Road"—a Japanese clone of the Harley. By 1937 gathering war clouds forced Harley-Davidson to abandon their Japanese links, but the Rikuo continued to prosper. Indeed, the Harley clone not only became the mainstay of the Japanese motorcycle industry, but was, ironically, also a valuable asset to Japan's Imperial Army, with more than 18,000 of the Rikuo Type 97 produced for military service.

An early Norton bike on view at London's Olympia Motorcycle Show in 1925.

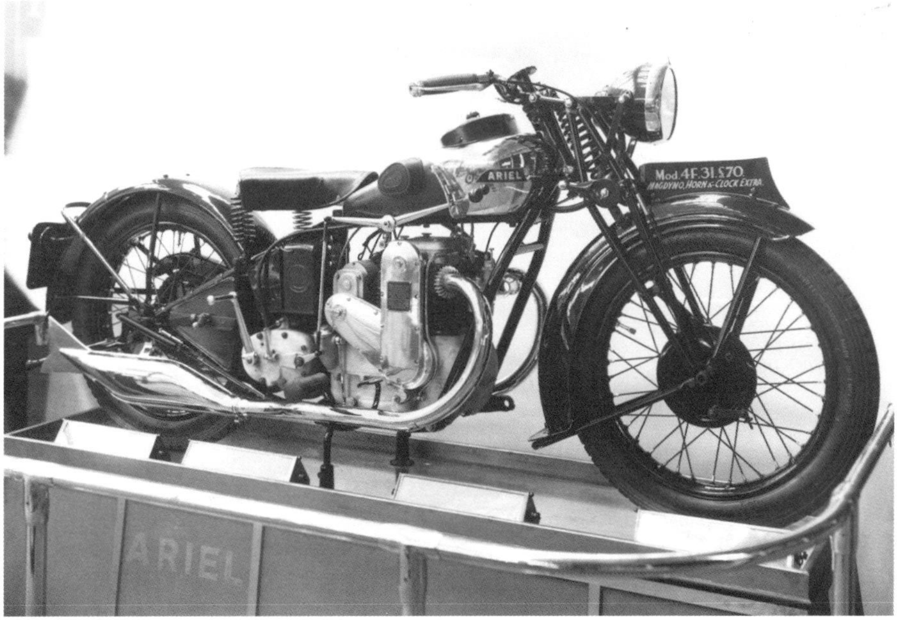

A four-cylinder Ariel displayed at the Olympia Motorcycle Show in 1930—a good year for British bike production.

Combat Bikes in World War Two

During World War Two, motorcycles were enlisted in military campaigns throughout the world. An effective and relatively inexpensive way to increase the mobility of fighting units, motorcycles fulfilled the role later taken on by armored personnel carriers and helicopters.

The German army in particular made expert use of the motorcycle. The Wehrmacht's access to tanks and armored vehicles was severely restricted by the Treaty of Versailles so, when Germany began rearming during the 1930s, the army turned to motorcycles. Bikes were ideally suited to the new kind of high-speed, "blitzkrieg" (lightning war) tactics developed by the German military. The Nazi invasions of Czechoslovakia, Poland, and France relied on speed, power, and mobility—and the motorcycle was uniquely able to meet these demands.

The combat success of Germany's motorcycle-equipped rifle regiments was not lost on Allied strategists. But the Allies never came up with a bike that could match the reliability, speed, or load-bearing capacity of the German army's Zündapp and BMW machines. Nevertheless, the Allies still produced some worthy bikes. In addition to supplying Allied armies with rifles and munitions, the British BSA company produced more than 120,000 of the M20, a formidable 500cc, side-valve motorcycle.

Uncle Sam's Two-Wheeled Warriors

In America 1943, the US War Department issued contracts to both Harley-Davidson and Indian to develop prototypes for new military motorcycles. Indian produced a military version of the Scout (the Model 741) and an experimental, shaft-driven Model 841. In the long term, however, the war brought little reprieve to the company, whose financial position became increasingly beleaguered. Indian bikes were mostly shipped to Britain and Canada and, by the end of the war, Indian was left with just one civilian model in its line-up—the Chief—while its dealer network was weakened and its market share dropped steadily.

Mobile, rugged, and reliable, motorcycles were in the front line of combat during World War Two.

The Wehrmacht were masters of motorcycle warfare. The blitzkrieg strategy of the German High Command relied on battalions of two-wheeled rifle troops.

Harley-Davidson, in contrast, was strengthened by its wartime service, despite the deaths of key members of the team that had originally forged the company—Walter Davidson in 1942 and William Harley in 1943. William Davidson had died earlier, in 1937, though brother Arthur remained active in the company until his death, aged 69, in 1950.

The Harley WLA, a Flathead 45, was the US Army's favored bike throughout the war, and nearly 90,000 were produced for the military. The WLA was equipped with ammunition boxes and a holster for a Thompson sub-machinegun. But, in contrast to the Wehrmacht's two-wheeled tactical assaults, the US Army mainly used its bikes for reconnaissance and courier work. Nevertheless, the wartime prevalence of the WLA gave many GIs their first taste of a Harley, something many would remember when they returned to the States.

BLITZKRIEG BIKES

During the late 1930s the Nazi war machine relied on its motorcycles. Both Zündapp and BMW produced models for military use. The German High Command specified construction requirements for the bikes, calling for a driven sidecar wheel, a reverse gear, and a specialized gearbox for off-road use. To meet the specifications BMW came up with the R75, while Zündapp produced the KS 750. More than 18,000 KS 750s were produced between 1941 and 1944—the bike's technology making it one of the most advanced motorcycles of the period. BMW's R75 was equally impressive. Featuring a 750cc OHV engine, the bike could tow a load of more than 800 pounds (400 kg).

Runaway Roadsters: Motorcycles in Postwar America

The postwar period was boom-time for the American economy. But Indian motorcycles, already struggling before the war, found the 1940s and 1950s tough going. Frustrated with the company's fortunes, the Du Ponts sold Indian to Ralph B. Rogers, a millionaire industrialist, in November 1945.

Rogers, the new owner of Indian, had a bold vision. Indian, he reasoned, could continue to appeal to traditional riders with the big, trusty Chief. But, he argued, the company could also attract a whole new set of customers through the production of "friendlier," lighter machines. Hence 1948 saw the launch of a 440cc, parallel twin version of the Scout, together with a totally new, single-cylinder bike—the 220cc Arrow.

Indian's Final Sunset

In essence, Ralph Rogers' idea was sound, but in practice Indian's attempt to revolutionize American motorcycling was a disaster. Hastily engineered and rushed into production, the new bikes quickly earned a terrible reputation. As a consequence, the company's sales nose-dived. To bring essential capital into the firm, Rogers struck a deal with Englishman John Brockhouse. In return for much-needed finance, Brockhouse was allowed to use Indian dealers to distribute his own, British bikes—a line-up that included such marques as Douglas, Matchless, Norton, Royal Enfield, and Vincent. For a time the relationship endured, but in 1950 Brockhouse called in the debt and Rogers was removed as Indian's president. The company was broken into separate sales and manufacturing concerns, with the pieces ultimately sold to Associated Motorcycles Ltd. (AMC), the British parent company of Brockhouse's motorcycle stable.

In 1950 the Chief was relaunched as the 1300cc Blackhawk Chief, one of the most impressive bikes in Indian's history. But by 1954, the Blackhawk Chief had been replaced by a 700cc Royal Enfield model rejigged for the American market through the addition of wide handlebars, fatter tires, and an "Indian" logo painted on the gas tank. But by the late 1950s AMC had dispensed with the Indian trademark altogether, and the sun set on a piece of American motorcycle history.

The "British Invasion"

AMC's hijack of Indian was indicative of a wider "British Invasion." The war had left the British domestic market depressed, so as an alternative British motorcycle firms looked to export sales—especially to the US.

During the late 1940s a string of new speed records established British bikes in the American public's mind as never before. Outstanding was the Vincent Black Shadow, a 998cc V-twin that clocked over 150 mph (240 km/h) on the Bonneville Salt Flats in 1948. Brashly marketed as "The World's Fastest Standard Motorcycle," the Black Shadow became the standard-bearer for the ranks of British bikes, among them Triumphs, Nortons, and BSAs.

"Panheads," Hydra-Glides, and Sportsters

Despite the British challenge, the going remained good for Harley-Davidson. Full civilian output resumed in 1947 and the company's market-share grew as Indian's fortunes dipped. In 1947 production of the revered "Knucklehead" engine was discontinued, but it was replaced by new, more advanced 1000cc and 1200cc models. Topping the heads of the new engines was a one-piece, chrome-plated, skilletlike cover that gave the engine its enduring nickname—the "Panhead."

Harley also gave more attention to styling, with the addition of extra chrome trim and the launch, in 1949, of the 74FL Hydra-Glide. With its hydraulically damped, telescopic forks, the 74FL was the first Harley-Davidson aimed squarely at the touring market. Continuing the trend, 1958 saw the launch of the Duo-Glide, whose rear hydraulic suspension provided an unbeatably smooth ride. Harley-Davidson also introduced some leaner, faster machines to give the British a run for their money. The 1957 XL Sportster, with its 900cc V-twin engine, was a nifty roadster that won a firm following among riders looking for speed and agility.

Admiring looks for the Vincent Black Lightning, on display in 1949, a year after it tipped 150 mph (240 km/h) at the Bonneville Salt Flats.

Far left: At the end of the war thousands of army surplus bikes provided cut-price access to the open road.

The First Wild Ones

The clean-cut motorcycle clubs that sprung up during the 1920s and 1930s continued to thrive during the 1940s and 1950s. Club enthusiasts met to shoot the breeze in local bars or clubhouses, and sometimes sported matching sweaters or jackets with their club logo—or "colors"—embroidered on the back. A member's social calendar often revolved around club road trips and get-togethers, while die-hard riders enjoyed testing their bike skills in organized track races and hill-climbing competitions.

Early Outlaw Clubs

On the fringes of the mainstream motorcycle scene other, less-respectable, groups developed. Their ranks included a few rebellious teenagers, but mostly they were loose fraternities of war veterans searching for camaraderie and excitement as they struggled to adapt to civilian life. Many had picked up technical expertise during the war and enjoyed tinkering with engines and pushing their bikes to the limit. But they were also a restless, disillusioned breed, and had a taste for marathon drinking sessions. Ostracized by the AMA for their throttle-happy, beer-fueled escapades, the outcast status of these groups earnt them a reputation as "outlaw" clubs.

The outlaws' motorcycles also had an aura of rebellion. Dissatisfied with heavy, accessory-laden production bikes, many outlaw riders stripped down their machines to the bare essentials. The fenders were "bobbed" (either shortened or removed altogether) and any other unnecessary trappings (saddlebags, windshields, chrome trim, big headlights) were chopped away. Some "bob-jobs" even had their heavy stock seats replaced by a tiny saddle bolted onto the frame. The idea was to improve the bike's performance through reducing its weight and drag. But "bobbing" also became an expression of attitude. A "bobbed" bike showed contempt for mainstream conventions and proclaimed the independence and individuality of its rider.

The Galloping Gooses and the POBOBs

It was common for many of the early outlaw clubs (they have never identified themselves as "gangs") to underscore their sense of alienation and nihilism through their choice of name. The Galloping Gooses, for example, were a motley group of ex-servicemen who took their name from the phrase they used for "giving the finger" (or "flipping the bird," the timeless gesture of defiant contempt).

Many of the first outlaw clubs sprang up around the dilapidated industrial towns of Southern California. San Bernardino

(or "Berdoo"), a blue-collar city 60 miles (100 kilometers) east of Los Angeles, was especially known for its active outlaw scene. Just south of Berdoo, the small town of Bloomington was home to a club who occupy a sacrosanct place in the folklore of outlaw bikers. Little is known of them for sure, but they called themselves the Pissed Off Bastards of Bloomington—or POBOBs.

"Wino" Willie and The Boozefighters

Los Angeles had equally wild characters. "Wino" Willie Forkner was especially notorious. During the war, Wino Willie had served in the US Army Air Corps. Discharged in 1945, he returned to Los Angeles and hung out with the 13 Rebels, a motorcycle club he had ridden with prewar. But in 1946 Willie's relations with the 13 Rebels cooled after he literally gate-crashed one of its race meetings by smashing his Indian Chief through the track fence and careering around the circuit at breakneck speed.

Shunned by the 13 Rebels, Wino Willie looked around for other bikers who shared his manic sense of fun. Sitting in southwest LA's All-American Bar with his buddies, Willie began kicking around possible names for a new club. "The Bats" and "Henchmen" were suggested, but when a local barfly suggested "The Boozefighters," Wino Willie and his partners all agreed the name was ideal.

The Boozefighters MC (Motorcycle Club) adopted a distinctive uniform of a football shirt with the club name emblazoned across the back. The Boozefighters, though always ready for a party, also took their riding seriously and sponsored field meets, runs, and races. In little more than a year, two more club branches ("chapters") had been added to the original club. With their love of madcap stunts and mayhem, the Boozefighters were on the margins of "official" motorcycle culture. The AMA refused to charter the club's name and abhorred its reputation for boisterous carousing—a notoriety that became even worse in 1947, after the events that unfolded at Hollister.

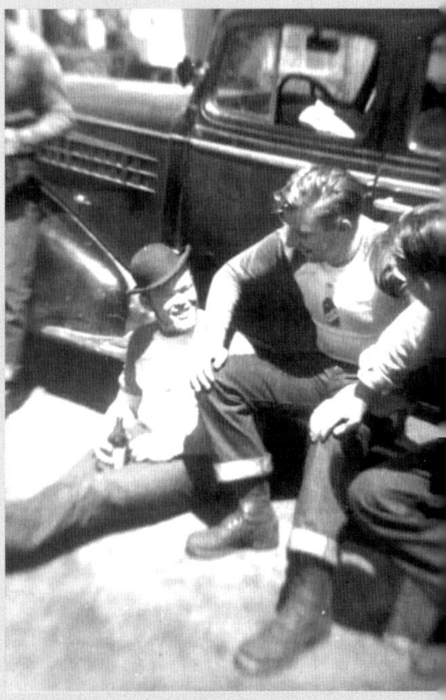

Members of the Boozefighters take some time out at the "birthplace" of the American biker—Hollister, California, July 4th, 1947.

Left: The Boozefighters were in the thick of the action at Hollister in 1947.

Far left: The legendary Boozefighters Motorcycle Club in the back room of the bar where it all began—the All American Bar, in South Gate, California, 1946.

Hollister: The Reality and the Myth

The sleepy Californian town of Hollister is roughly 95 miles (155 kilometers) south of San Francisco. By the 1940s it already had strong motorcycle connections, hosting its first Gypsy Tour in 1936. Organized by the AMA since the 1920s, Gypsy Tours were long-range rides to motorcycle rallies, and were known as friendly, good-humored events. The town even boasted its own collection of biker roughnecks, The Top Hatters Motorcycle Club.

In 1947 few people living in Hollister worried when plans were announced for a Gypsy Tour to converge on their town over the Fourth of July weekend. Sanctioned by the AMA and sponsored by the Salinas Ramblers Motorcycle Club and the Hollister Veterans' Memorial Park Association, the event was meant to be a relaxed weekend of racing and innocent recreation. But what ensued became infamous.

"We're Just Having a Convention"

A Gypsy Tour hadn't been held anywhere in California since before the war, and word quickly spread that the event would be a blast. Clubs like the POBOBs and the Boozefighters weren't going to be left out, and were soon heading to Hollister. By Friday night about 2,000 bikers had arrived, the crowd probably peaking at around 4,000.

The vast majority of the visitors camped peacefully on the edge of town. But a more raucous group of maybe 500 bikers congregated in the bars lining San Benito Street, Hollister's main thoroughfare. With the Boozefighters and POBOBs taking the lead, the event turned into an uproarious, drunken binge. San Benito Street became an impromptu dragstrip, with bikes thundering up and down town. Inebriated riders attempted risky stunts to entertain the raucous throng, and Hollister's half-dozen cops had little hope of maintaining order. "The American Legion goes into town and raises hell," a young rider told newspaper reporters. "It's a convention," he explained, "We're just having a convention."

As the partying continued, a special court session had to run through Saturday night to deal with the number of arrests. "If we jailed everyone who deserved it, we'd have herded them in by the hundreds," one exasperated cop complained. In total, 49 arrests were made, with $2,000 worth of fines and an undetermined number of jail sentences passed for drunkenness, disturbing the peace, and reckless driving. Among those convicted, Frisco Boozefighter "Kokomo" McKell was fined $25 and given two days in jail for disturbing the peace, while a young LA Boozefighter—19-year-old Jim Morrison (not the Jim Morrison of The Doors)—was given 90 days in the county slammer for indecent exposure.

Order was finally restored on Sunday night when the town cops were reinforced by a 40-man squad from the California Highway Patrol. Armed with tear gas and riot gear, they shepherded the bikers into one section of San Benito Street. To preoccupy the crowd, a dance band (originally booked for a show at the American Legion Hall) was ordered onto the back of a nearby truck to play a few numbers. Enjoying what was left of the weekend, the bikers bopped their way through the inches of broken glass—debris from the previous day's bottle barrages. By Monday morning it was all over and the bleary-eyed riders gunned their bikes out of town.

Havoc at Hollister

Initially, Hollister wasn't a huge news story. The town had seen a wild weekend, but it had hardly been devastated. Arrests were mainly for misdemeanors, mostly drunkenness and disorderly conduct. And, while 40 people were treated at the local hospital, the only serious injuries were sustained by three bikers hurt during their own drunken stunts. In fact, the whole brouhaha would probably have been quickly forgotten had it not been for photographer Barney Peterson.

Peterson had turned up in Hollister to snap some shots of the Gypsy Tour. Missing most of the bike action, he got a great picture by persuading a passing drunk to pose, slumped on a parked

Far left and right: In 1947 the streets of Hollister were host to a wild weekend that has become part of biker folklore.

motorcycle. Nearly three weeks later, the shot was splashed across a whole page of *Life*, America's leading news magazine. In the accompanying story, a drunken spree by a few hundred bikers was transformed into a merciless invasion by an army of 4,000. Readers were left with the abiding impression of a small-town community plunged into total chaos by a legion of two-wheeled savages.

The Myth of the Motorcycle Menace

Picked up by the *New York Times* and the *Los Angeles Times*, the story developed a life of its own. Hollister became fixed in the public mind as the town laid waste by a barbarian horde. The image was replayed endlessly in the media. In 1951, for example, a fictionalized version of the Hollister incident appeared in "Cyclists' Raid," a short story published in the popular *Harper's* magazine. The author, Frank Rooney, was influenced by the larger-than-life press reports of Hollister, his story depicting an unnamed American town impotent before a marauding motorcycle gang. In Rooney's overwrought melodrama, the local sheriff is powerless to stop the predatory bikers and, when a young girl is killed, her father is helpless against the regimented, faceless motorcycle pack.

This representation of the biker as a menacing "bogeyman" became a recurring media stereotype. Replayed in endless books, movies, and magazines, the image of a feral motorcycle gang rampaging through Main Street, USA, is an enduring image. It was a myth that would affect the public's perception of bikers for many years to come.

***Life* magazine's coverage of the Hollister "riot" in 1947. Strictly speaking, the photo wasn't faked, but it was certainly highly staged.**

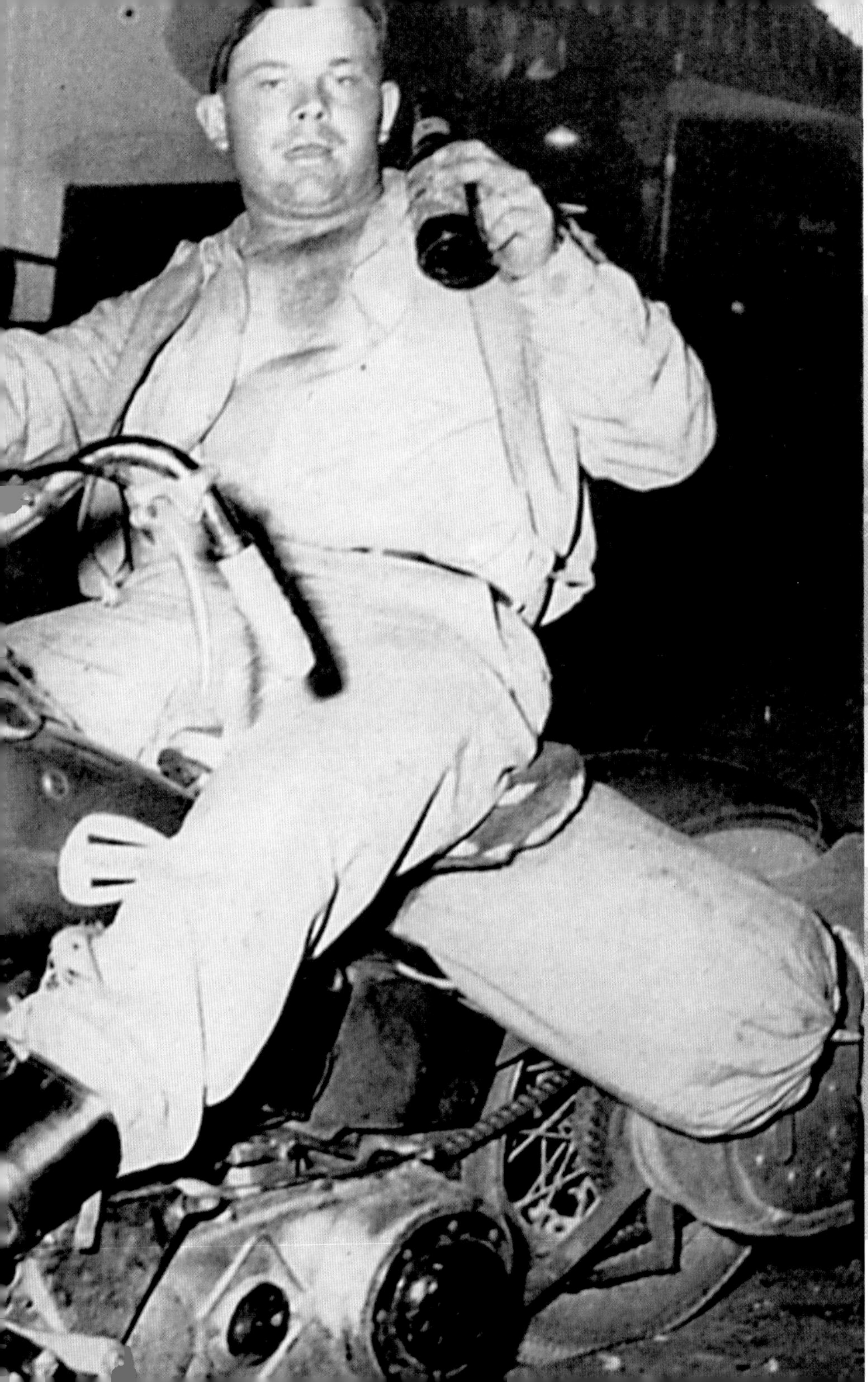

THE PHOTO THAT SPAWNED A LEGEND

Barney Peterson's Hollister photo appeared in *Life* magazine on July 21st, 1947. Strictly speaking, the picture wasn't a fake, but it involved a big measure of theatrical license. Posing outside Johnny's Bar and Grille on San Benito Street, a tanked-up biker on a scruffy motorcycle (a "bobbed" Harley EL) waves a beer bottle in each fist and rests his jack-booted foot above a heap of broken glass. The fact that Peterson's own newspaper—the *San Francisco Chronicle*—didn't run the photo, suggests that locals didn't consider Hollister a major story. But, after the image was published nationally in *Life*, the Hollister events were transformed into headline news. The notoriety of outlaw bikers had begun.

THE BIRTH OF THE "1% ERS"

Anxious to disassociate itself from the Hollister ruckus, the American Motorcycle Association reputedly insisted that 99 percent of American motorcyclists were respectable, well-behaved citizens—and only a small minority were troublemakers. Others, however, reveled in the infamy. Ever since Hollister, the "1% er" designation has been enthusiastically embraced by "Outlaw" bikers. Taken as a badge of honor, the "1%" patch is worn proudly on the jackets of all outlaw club members.

"The Wild One"

In 1954 *The Wild One* took the Hollister myth to the big screen. The movie's plot mimicked the Hollister press stories. After being thrown out of a motorcycle meet, a bike gang—the Black Rebels Motorcycle Club—"invade" a small town, get drunk, and stage their own event. The gang leader, Johnny Strabler, falls for the local cop's daughter, but things turn nasty when a rival gang—the Beetles—roll into town.

"Hey Johnny, What're You Rebelling Against?"

To direct the picture Columbia Picture's producer Stanley Kramer recruited Laslo Benedek, an erudite Hungarian, while the script (based on Frank Rooney's short story) was provided by John Paxton. Marlon Brando was signed up to star as Johnny Strabler, and turned in a classic performance: a study in taciturn brooding, encapsulated in a classic snatch of dialog. "Hey Johnny, what're you rebelling against?" asks a pretty local girl. "What've ya got?" growls back the leather-jacketed gang leader. Loaded with such defiant nihilism, Brando's character crystallized the biker's rebellious mystique.

Kramer was eager to give *The Wild One* an edge of gritty authenticity. He wanted to shoot the movie in Hollister but Columbia Pictures' president, Harry Cohn, insisted it was made on his studio's Burbank backlot. Real-life bikers were recruited as extras, and Kramer claimed snatches of their conversation were put into the movie's dialog. Brando's legendary retort, "What've ya got?," was supposedly taken from a surly biker. Other lines, however, were less convincing—one of the Black Rebels yelling to a doddery barman, "Say Pops, gimme another one of those crazy beers, will ya, man."

The Legacy of "The Wild One"

On its release, *The Wild One* was dogged by controversy. Studio boss Harry Cohn hated the picture and, amid a climate of unease about juvenile crime, many newspapers and magazines slammed the movie as a celebration of delinquency. In Europe, too, *The Wild One* was a target for moral crusaders. In Britain the British Board of Film Censors, fearing the movie would incite trouble, banned public screenings for 14 years.

Despite being abhorred by officialdom—indeed, partly because it was abhorred by officialdom—*The Wild One* was a hit with rebellious movie-goers. Young audiences especially were enamored of the picture's seditious cool, and cheered on Brando and his motorcycle cohorts. And, while the original Boozefighters were lukewarm toward the movie, *The Wild One* has been revered by subsequent generations of bikers. During the early 1960s Frank Sadilek, president of the Hell's Angels' San Francisco chapter, even bought the striped shirt Lee Marvin wore in the movie, and made a point of wearing it whenever he had to meet the police.

The 1954 movie *The Wild One* gave a Hollywood spin to the "havoc" at Hollister.

MARLON BRANDO

Before appearing in The Wild One, *Brando was already an established star, having wowed audiences in 1951's* A Streetcar Named Desire. *Brando agreed to star in* The Wild One *because he admired Kramer's work, and because the producer had given Brando his first big break in the 1950 movie,* The Men.

Brando hit it off with the bikers hired as movie extras, and incorporated their mannerisms in his characterization. Brando also selected his own wardrobe for the movie, but—although he owned his own motorcycle—he struggled to master the 650cc Triumph Thunderbird he rides in the movie.

LEE MARVIN

In The Wild One, *Lee Marvin made his movie debut as Johnny Strabler's nemesis—Chino, the leader of the Beetles motorcycle gang. An actor for only three years, Marvin had been a marine during World War Two and was seriously wounded on the beaches of Saipan. Chewing on a cheroot and spoiling for a fight, Marvin played Chino as a wild screwball. He reputedly based the part on the notorious Boozefighter, "Wino" Willie Forkner. Marvin had met "Wino" Willie while preparing for his movie role in Los Angeles. The fact that Marvin loathed Marlon Brando also added extra tension to the set—and ensured the fistfight between their movie characters has exceptional realism.*

STANLEY KRAMER

One of Hollywood's first independent producers, Stanley Kramer had a reputation as a liberal whose pictures had a social conscience. He had the idea for The Wild One *after stumbling across Frank Rooney's version of the Hollister saga in a copy of* Harper's *magazine.*

Kramer's original intention was to give The Wild One *a social message. In Kramer's proposed ending the town merchants refuse to press charges against a marauding motorcycle gang because of the dollars the bikers have pumped into their coffers. During the Cold War era, however, film censors found this critique of capitalist hypocrisy too contentious. As a consequence, the movie's liberal moralizing was downplayed in favor of an enigmatic romance between gang leader Johnny Strabler (Marlon Brando) and archetypal good girl, Kathie Bleeker (Mary Murphy).*

The Black Leather Look

In appearance, the two gangs of *The Wild One* were very different. Unkempt and dressed in an assortment of tatty denim and old army fatigues, Chino's Beetles were probably closest to the appearance of outlaw bikers of the day. Togged up in designer leathers (Brando sported a Perfecto "One Star" jacket made by Schott) and blue jeans rolled up above their bike boots, Johnny and the Black Rebels were more of a stylized Hollywood ideal of what a motorcycle gang *should* look like.

The Earliest Motorcycle Outfits

During the early twentieth century the first motorcycle riders wore long, duster-style coats for protection against both the weather and messy bike engines. Alternatively, they adopted the coats and boots sported by aviators who, in turn, had adapted the wear of cavalry officers. There was a particular preference for leather coats and jackets—not only because they offered protection from the elements, but also because leather provided a tough second "skin" that guarded against scrapes and grazes.

The First Biker Leathers

The first coats designed specifically for motorcyclists appeared in about 1910. Mid- or full-length coats, they were generally made from black horsehide or pigskin, and they borrowed their design from men's raincoats. Early pioneers of motorcycle-wear were Irving and Jack Schott, the sons of Russian immigrants, who opened a workshop on Manhattan's Lower East side in 1913. Originally makers of raincoats, the Schott brothers began producing motorcycle jackets during the 1920s. Leathertogs in Boston were another pioneer brand, producing leather garments for Harley-Davidson as early as the 1910s and 1920s.

In Britain, 1899 saw Lewis Leathers start life as a tailors, D. Lewis and Sons, on London's Great Portland Street. In 1926 Lewis launched a leather coat business under the "Aviakit" label. Its leathers were originally intended for pilots, but by the 1930s many army bikers and some racers were wearing Lewis jackets. In the 1950s and 1960s the company become a favorite among Britain's café racers, and its products also enjoyed a good reputation abroad.

The Classic Black Leather Jacket

During the 1930s and 1940s the US military developed many styles of zippered, leather flight jackets. Well-liked by soldiers and aircrew, many jackets were still worn after their owners had returned to civilian life. These "bomber" jackets became especially popular for motorcycle riding.

The commercial production of motorcycle jackets shifted into top gear during the 1940s and 1950s. Flight jacket design was still a big influence, but new motorcycle styles also began to appear. Schott's "One Star" and Harley-Davidson's "Cycle-Champ" ("Cycle Queen" for women) were typical of the new "W"-style jackets that became fashionable. With an angled front zipper and an overlapping "W"-shaped collar, this was the design popularized by Marlon Brando in *The Wild One* and, as a result of the movie's influence, it was the style that became recognized as the familiar black leather jacket of the biker.

Developed for flight crew during World War Two, the classic leather "bomber" jacket later became a firm favorite among bikers.

02

HELL'S HIGHWAY TO HOLLYWOOD

Outlaws on the Silver Screen

Screening the Asphalt Cowboy

Popular culture has always been beguiled by the mystique of the motorcyclist. The wild outlaw, barreling down the highway astride a two-wheeled mustang, is an image of reckless freedom that the media has found irresistible.

But media representations of the biker have always been double-edged. On one hand, he has been demonized as the antithesis of civilized society, presented as a marauding barbarian, hell-bent on an orgy of rape and pillage. On the other hand, the "bad boy" biker has also been cast as an embodiment of individuality and raw adventure. Here, the shudders of fear give way to a frisson of excitement, the biker promoted as a charismatic renegade who defies the suffocating conventions of the workaday world. From this perspective the outlaw biker is like a modern-day Daniel Boone—an asphalt cowboy who is the last remaining heir to America's original frontier spirit.

At the movies, especially, the biker has been recurrently depicted as an enigmatic rebel. The silver screen outlaw may be a snarling outsider, but his dangerous élan has had strong appeal with audiences seeking the vicarious thrills of defiant cool. This electrifying edge is what made *The Wild One* a two-wheeled tour-de-force. It was also a quality that Hollywood's glut of 1960s "chopper operas" had in spades. During the late 1950s and early 1960s, however, the celluloid "motorcycle menace" looked decidedly less threatening.

The *Beach Party* films of the early 1960s featured the "Rats and Mice"—a gang of bumbling bikers.

The Mild Ones

Despite its enduring impact, *The Wild One* spawned few immediate imitators. During the 1950s the Brando classic was just one among many motion pictures prompted by contemporary fears of growing juvenile delinquency. In 1955, for example, the *The Wild One* was followed by Warner's *Rebel Without a Cause* and MGM's *The Blackboard Jungle*. Produced by major studios, these movies were sobering portraits of dysfunctional adolescence. But lower-budget fare was more equivocal. Pictures such as *Untamed Youth* (1957) and *High School Confidential* (1958), for instance, purported to preach against the "evils" of juvenile crime but also offered teen audiences the buzz of delinquent rebellion.

The number of these "teenpics" multiplied as Hollywood, facing a decline in adult audiences, turned its attention to the youth market. Movies such as *Rock Around the Clock* (1956) capitalized on the rock 'n' roll boom, while the latest teenage fads were exploited in movies like *Hot Rod Gang* (1958) and *Ghost of Dragstrip Hollow* (1959). The field was led by a new, independent studio, American-International Pictures (AIP), whose movies were perfectly tailored to the growing teenage demand. In AIP's approach to moviemaking, every expense was spared—the studio cranking out pictures as quickly and cheaply as possible. But, despite their bargain-basement production values, AIP's movies were always rammed-full of the thrills, spills, and sensation that young audiences clamored for.

Released in 1957, *Motorcycle Gang* was AIP's attempt to emulate the volatile tension of the *The Wild One*. The plot of *Motorcycle Gang* focused on the rivalry between a reformed delinquent (played by Steve Terrell) and a biker firebrand (John Ashey). The movie featured some neat Triumph and BSA scrambling action from The Skyriders motorcycle club, and the picture's leather-jacketed bad boys were suitably mean. But *Motorcycle Gang* was short of the noir-esque elements that gave *The Wild One* its intensity. Filmed in bright sunshine, *Motorcycle Gang*'s cast looked disconcertingly wholesome, while lame back-projection, slapstick, and wisecracks ensured the movie's thrill-factor rarely shifted out of first gear.

Hoodlum bikers put in a more convincing appearance in AIP's *Dragstrip Riot* (1958). A short (only 68 minutes), punchy teenpic, *Dragstrip Riot* chronicled the hell that breaks loose when a gang of motorcycle punks square up to a bunch of greasy hot rodders. *Dragstrip Riot* featured some great hot-rod footage and a teen-pleasing, enduring rock 'n' roll anthem—"Teenage Rumble"—but the movie was to be the last gritty motorcycle flick of the decade. By the end of the 1950s, the image of the moody biker was out of step with the direction of AIP's movie releases. A very different take on the motorcycle mythology was beginning to take its place.

The publicity for *Motorcycle Gang* promised a wild and wicked ride—but the film itself barely opened the throttle.

A teenpic classic, *Dragstrip Riot* saw greasy hot rodders duke it out with badass motorcycle punks.

Beach Blanket Bikers

During the early 1960s, AIP's output was dominated by a succession of beach-bound comedies. Beginning with *Beach Party* in 1963, the AIP beach movies were a bubbly blend of music and gags that centered around a group of chirpy Californian teens, led by pop heartthrob Frankie Avalon and Annette Funicello (a former Disney mousketeer). But, while the spotlight of AIP's beach movies was on the surfside hijinks of Avalon, Funicello, and friends, the series also regularly featured a gang of bumbling bikers.

The Rats and Mice motorcycle gang of AIP's beach movies was a comic foil to the surfers' playful antics. Led by the patently middle-aged Eric Von Zipper (played by Harvey Lembeck, who mercilessly sent up Brando's smoldering rage), the blundering *Beach Party* bikers were a parody of the 1950s wild ones. In movies such as *Bikini Beach* (1964), *Pajama Party* (also 1964), and *Beach Blanket Bingo* (1965), the Rats and Mice gang featured as an emblem of outdated rebellion.

For a while, then, it seemed like bikers were old news. By the early 1960s the good times were rolling—JFK was President, the US economy was booming, and sun-kissed, teenage fun was the order of the day. But, while tinseltown lampooned the surly biker as an anachronistic joke, out on the streets the outlaws were mustering for action.

Bikini Beach **featured more wacky comedy from Eric Von Zipper and his goofball gang.**

AMERICAN INTERNATIONAL PICTURES

American International Pictures (AIP) was the indisputable teenpic king. The studio was founded in 1954 by James H. Nicholson (a clued-up movie executive) and Samuel Z. Arkoff (a cigar-chewing showbiz lawyer). The trailblazing partners were the first to recognize the huge sales potential of a teenage audience, and over the next 30 years AIP bombarded American teens with a relentless flood of comedy, action, and horror.

During the 1950s AIP built its reputation on hurriedly made, low-budget pictures such as *Reform School Girl* and *I Was a Teenage Werewolf* (both 1957) —movies that pitted wayward teenagers against figures of conservative authority. AIP was also master of the hard sell. Sometimes Nicholson and Arkoff would come up with a gripping title and a startling sales campaign long before giving any thought to a picture's actual content.

The Beast is Born: The Rise of the Hell's Angels

After 1947's debacle at Hollister, it was business as usual for the Boozefighters. In 1948 the city of Riverside had been chosen as the site for another Gypsy Tour and all three Boozefighter chapters went along for the fun. Some members even dyed their hair bright red to make an especially outrageous impression.

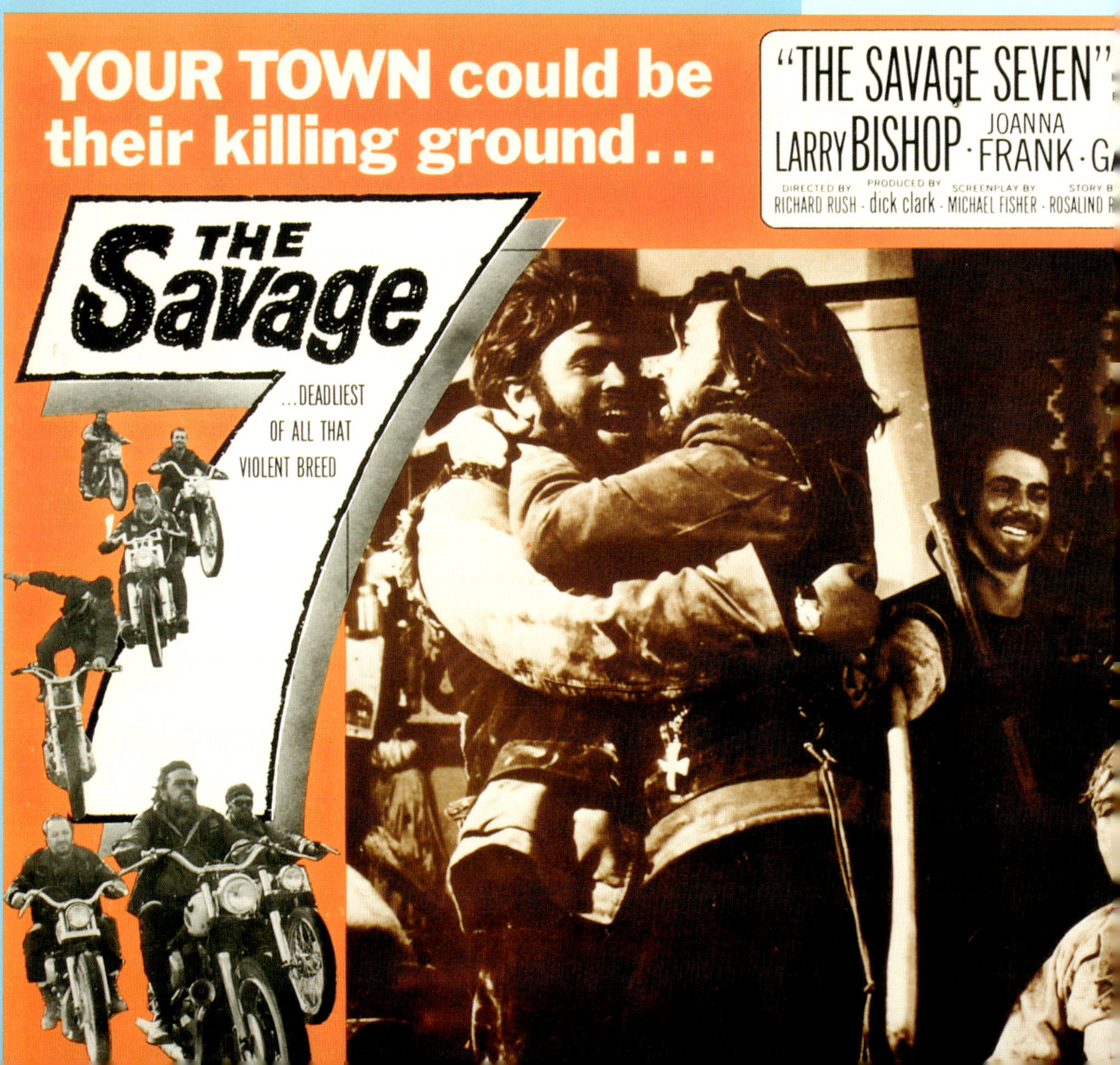

After the "invasion" of Hollister in 1947, the menace of outlaw motorcycle gangs in small-town America became a stock cliché in low-budget exploitation movies.

Just like Hollister, the Riverside event (sponsored, somewhat ironically, by the Sheriff's Training Association) quickly degenerated into a wild spectacle of impromptu drag races and ceaseless beer-guzzling. This time, however, the local newspapers paid close attention, running pictures of Boozefighters "Fat Boy" Nelson and Jim Cameron astride their bikes, showing off their club sweaters and cheerfully chugging back beer. And, although no Boozefighters were actually arrested at Riverside, the press didn't hesitate to condemn the club as prime ringleaders of the trouble.

Throughout the 1950s the Boozefighters maintained a tough reputation, but by the end of the decade it was time for a new generation of motorcycle mavericks to take centerstage.

Angels Let Loose

Precisely how the Hell's Angels Motorcycle Club came to be formed is shrouded in a haze of burnt rubber and exhaust fumes. But there's general agreement that the club developed from the remnants of the legendary Californian outlaws, the POBOBs. By 1948 the POBOBs had splintered, with some members drifting to Fontana in San Bernardino County. In March 1948 they established a new club, and took a name that would become infamous—the Hell's Angels.

Back in the 1920s, an AMA-sanctioned club known as the Hell's Angels had existed in Detroit, but they had no link to the later outlaws. It's more likely that the southern Californian bikers hit on "Hell's Angels" through the name's wartime association with a variety of fearless flyboys.

By 1954 a San Bernardino Hell's Angel remembered only as "Rocky" had wandered north. Winding up in San Francisco, Rocky launched a second incarnation of the club. The newly established San Francisco Hell's Angels drew much of its original membership from a local biker outfit, the Market Street Commandos (Market Street being one of San Francisco's more riotous neighborhoods). From the mid-1950s to the early 1960s the Frisco Angels were headed by Frank Sadilek, a charismatic organizer usually credited with designing the club's official "Death's Head" insignia—a grinning skull wearing a pilot's helmet with extended wings. Like the club's name, the Hell's Angels' insignia also had military derivations, probably inspired by the "winged skull" emblems used by the 85th Fighter Squadron and the 552nd Medium Bomber Squadron. The club's trademark colors—red and white—were also inspired by Air Force insignia.

"Mother" Miles and "Mighty Mouse"

A third Hell's Angels group was established in North Sacramento. The prime movers were James "Mother" Miles and his brother, Pat. At 5'10" (1.78 m), and weighing around 240 pounds (109 kg), "Mother" Miles was a mountain of a man. But his nickname came from his benevolent nature. "Mother" was always happy to bail fellow riders out of jail, or give them a roof for the night. Much smaller than his brother, Pat Miles was known as "Mighty Mouse." As teenagers during the mid-1950s, they had formed the Hellbent for Glory motorcycle club but bigger things were on the horizon. Joining with Don "Boots" Reeves (a local biker and aspiring Country singer), Mother Miles and Mighty Mouse revamped Hellbent for Glory as a North Sacramento branch of the Hell's Angels.

At this stage, however, the various Hell's Angels chapters still operated independently. Loosely organized and unstructured, there were few close links between the different factions.

The Beast is Born: The Rise of the Hell's Angels

Hell's Angels in God's Country

Life wasn't easy for the North Sacramento Hell's Angels. Continuously harassed by the local police, Mother Miles and most of the club headed out of town, starting a Hell's Angels "Nomad" chapter based around Richmond, just across the Bay from San Francisco. Laying up in the Bay Area, Miles and his companions found they got less aggravation from the cops—and for that reason the Bay Area soon became known as "God's Country."

Across the Bay Bridge from San Francisco there also lay the blue-collar city of Oakland. Here, another chapter of the Hell's Angels was established in 1957 by "Boots" Reeves (who had drifted down from Sacramento) and several others, including a young biker by the name of Ralph "Sonny" Barger. And, after Boots split from the club, it was Sonny Barger who took command.

With Barger's skillful leadership, the Oakland Hell's Angels grew in stature. Relations between the various chapters of the club were occasionally fractious but, under Barger's aegis, they gradually came together. Greater structure and organization developed, together with a system of club bylaws, codes of conduct, and chain of command. Taking pride in their defiant non-conformity, the Hell's Angels also developed a uniquely provocative style that combined long hair, Nazi motifs, greasy jeans, and stripped-down, mean-looking bikes. Nicknames also became a Hell's Angels' trait, and helped develop a sense of fraternal unity. The Oakland chapter, for example, boasted such oddball aliases as Terry the Tramp, Tiny, Zorro, Magoo, Fat Freddie, and Moldy Marvin.

Early bikers took pride in their defiant nonconformity and fraternal loyalty.

Named "Hell's Angels" by its crew, a B-17 attached to the 303rd Bombardment Group notched up 48 successful combat missions before being retired from service.

Death's Head Incorporated

The Hell's Angels' insignia also grew in importance. Since the 1920s motorcycle clubs had sported their own badges, but the Hell's Angels regarded its club "colors" as sacred. Embroidered on the back of a leather or cut-off denim jacket, the Angels' colors placed the club's "Death's Head" motif in the center, with "Hells Angels M.C." emblazoned across a top patch ("rocker"), while a bottom patch identified the location of the member's chapter.

By the late 1960s the Hell's Angels were growing into a slick operation. It became especially hard-nosed in enforcing ownership of its name and insignia, and by 1967 had registered itself as the "Hells Angels Motorcycle Club, Inc." Even their Death's Head insignia was trademarked, along with the name "Hells Angels" (the apostrophe being dropped in the process).

Forging themselves into a solid brotherhood, the Hell's Angels also became more selective about who they accepted as members. The practice of "prospecting" developed, whereby a club member could recommend someone for membership, but before acceptance the candidate would have to serve as a "Prospect"—doing club members' bidding (no matter how demanding or ludicrous) until he was officially approved as a "full patch" Angel.

Over the years the organization of the Hell's Angels has been widely emulated. During the late 1950s and early 1960s, however, it was still just one of many biker clubs alongside the Gypsy Jokers, the Hangmen, the Coffin Cheaters, and many more. But it all changed in the mid-1960s, when the Angels made headline news.

THE "HELL'S ANGELS" LEGACY

The name "Hell's Angels" springs from a long military lineage. Originally a moniker used by flying aces during World War One, the name was subsequently used as the title for Howard Hughes' 1932 air war melodrama starring Jean Harlow and Ben Lyon.

Before Japan's attack on Pearl Harbor in 1941, a squadron of the "Flying Tigers"—the covert US Army Volunteer Group that flew fighters in support of China—was also called "Hell's Angels." During World War Two, numerous B-17 and B-26 bombers were also given the name. One of the most famous was a B-17 attached to the 303rd Bombardment Group stationed at Molesworth in England. The plane flew no less than 48 successful combat missions, and in tribute to the plane's achievements the whole 303rd was renamed "Hell's Angels" in 1944.

Although many other military detachments also adopted the name, none of these wartime units had a direct link with the motorcycle club founded in 1948.

RALPH "SONNY" BARGER

Ralph "Sonny" Barger was a teenage warehouseman when he helped establish the Oakland chapter of the Hell's Angels in 1957. Born in Modesto, California, in 1937, Barger was just six months old when his mother walked out on the family. Raised by his alcoholic father, Barger grew to resent authority and became a self-assured rebel. Quitting school, he lied about his age and joined the US Army. But the facts caught up with him 18 months later, and he was given an honorable discharge.

Back on the streets of Oakland, Barger found new direction as a motorcycle outlaw. He was pretty skinny for a biker—6 feet (1.8 m) tall and barely 150 pounds (68 kg)—but his talents lay in his shrewd intelligence and natural aptitude for leadership. As long-time president of the Oakland Hell's Angels, Barger oversaw the consolidation of the club across California during the 1960s, and its worldwide expansion during the 1970s.

The Menace Returns

In 1964 the Hell's Angels became notorious. That year the club's annual Labor Day run to Monterey ended in outrage. The Angels had set up camp around a bonfire built on the beach, and the usual carousing ensued—though this time events took a nasty twist. Early on Sunday morning deputies stumbled across two teenage girls near the campsite. Dazed and half naked, the girls alleged they had been raped by a group of bikers. The cops rounded up some Hell's Angels and charged Mother Miles, Terry the Tramp, Crazy Cross, and Moldy Marvin.

Media stories about "marauding Hell's Angels" prompted the police to become increasingly tough with outlaw riders.

The Angels protested their innocence. Witness testimonies were also confused, and the evidence was inconclusive, so within three weeks all charges were dropped. But the story had made headline news. California's State Senator, Fred S. Farr, was irate and instructed the state's new Attorney General, Thomas Lynch, to launch a full investigation into outlaw motorcycle clubs.

"No Act Is Too Degrading for the Pack"

Made public at a press conference in March 1965, the Lynch Report seemed to confirm the most lurid rumors about motorcycle outlaws. Cobbled together from stories trawled up from police files across California, the report presented a litany of assaults, rapes, sexual perversions, and drug abuse.

On closer reading, however, the Lynch Report was less convincing. Taken individually, most of the "hoodlum activities" amounted to little more than drunken punchups—few of which had even resulted in arrests. Overwrought and melodramatic, the report was determined to malign outlaw bikers—and the Hell's Angels in particular. The Attorney General's account seemed desperate for meaty copy, and resorted to the fatuous claim that "the most universal common denominator in identification of Hell's Angels is their generally filthy condition…both club members, and their female associates, seem badly in need of a bath."

Nevertheless, a voracious press devoured the Lynch Report's spurious revelations. Throughout 1965 and 1966 stories in magazines such as *Time*, *Life*, *Look*, *Newsweek*, and *Esquire* further cemented the outlaw biker's infamous reputation, with *Time* magazine ruefully concluding that "No act is too degrading for the pack."

After the publication of the Lynch Report in 1965, the *Saturday Evening Post* was just one among many magazines that clamored to "expose" the Hell's Angels "menace."

The Laconia "Riot" of '65

The press alarm surrounding belligerent motorcycle gangs gave focus to a broader climate of anxiety. By 1965 the US economy was faltering, and liberal optimism was crumbling amid racial violence, urban disorder, and spiraling social discontent. Against this backdrop, the outlaw motorcycle clubs were vilified as the dreadful epitome of the nation's turmoil and decline.

In response, policing of bikers became increasingly ruthless, with police often reacting with brutality at the slightest hint of trouble. In June 1965 matters came to a head at Laconia.

On the shores of Lake Winnipesaukee in New Hampshire, Laconia's Weir's Beach is the location for one of America's longest-running motorcycle rallies. Following the Lynch Report and the attendant media hullabaloo, however, the town authorities anticipated trouble at the 1965 meeting. Stricter public order laws were rushed through the courts while the National Guard practiced their riot control drills. Amid the weekend's atmosphere of tense expectation, a flare-up became inevitable. When some revelers eventually let off smoke bombs and firecrackers the law moved in and things turned ugly. Following several hours of ferocious fighting, 70 people were left injured (some badly) and the police had made 34 arrests.

Evil Incarnate

The Hell's Angels hadn't been involved in the battle. Laconia was 2,000 miles (3,200 km) from the Angels' home turf, and few club members had bothered making the trip. Nevertheless, the local mayor immediately blamed the Angels for the mayhem. For months, media uproar had convinced the public that outlaw bikers were evil incarnate, and that the Angels were the worst of the bunch.

Wild Angels

During the mid-1960s the Hell's Angels probably weren't any bigger—or any more debauched—than any number of other outlaw motorcycle clubs. The Gypsy Jokers, Satan's Slaves, and many more were up there too. But it was the Angels that the media seized upon, branding them as the meanest mothers in the valley. Consequently, an "elite" status was conferred on the club, and any move the Hell's Angels made prompted a media circus, even at moments of tragedy.

Legendary Hell's Angel, Jim "Mother" Miles was killed in January 1966. After meeting with other Angels in Oakland, Mother Miles had been riding home to San Leandro (having recently moved east of the Bay) when he was hit by a truck—and died in the hospital several days later. As one of the most illustrious figures in biker folklore, Mother Miles' Sacramento funeral was a mammoth event, honored by a motorcycle convoy hundreds strong. Predictably, the media also turned up, the occasion getting nationwide TV coverage and a big spread in *Life* magazine.

The *Life* feature caught the eye of the boys at American International Pictures. With the Vietnam War escalating and the inner cities burning, AIP's chirpy beach movies were looking distinctly incongruous. Instead, AIP was looking for a topic more in keeping with the national mood of apprehension. Outlaw bikers seemed ideal, and the studio quickly entrusted a new movie project to Roger Corman, one of AIP's principal directors.

"Their Credo is Violence, Their God is Hate ..."

With the tag-line, "Their credo is violence, their God is hate and they call themselves the Wild Angels," AIP's first "chopper opera" billed itself as "the most terrifying movie of our time." Aiming for stark authenticity, director Corman and scriptwriter Charles Griffith undertook research by hanging out with Californian bikers—and hired members of the Hells Angel's' Venice chapter as extras.

George Chakiris, who had previously played a gang leader in *West Side Story* (1961), was lined up to play a rogue biker, Jack Black, with Nancy Sinatra costarring as a sultry love interest. Shortly before shooting started, however, it was discovered that Chakiris couldn't ride a motorcycle and, after a wobbly tryout,

In *The Wild Angels* Peter Fonda plays the leader of a motley gang of outlaw bikers.

the actor declined to learn. Loath to use a stunt double, Corman replaced Chakiris with Peter Fonda, who agreed to take the part on condition the character's name was changed to Heavenly Blues (the street name for a narcotic made from morning glory seeds).

Filmed in just 15 days on a spartan budget of $360,000, *The Wild Angels* exemplified Corman's "quick and dirty" oeuvre. As Blues, Fonda plays the leader of a tough biker gang hunting down a stolen motorcycle. After a confrontation in Mexico, gang member "Loser" (Bruce Dern) is fatally wounded and his fellow Angels resolve to return his body to his hometown for burial. The movie culminates in an orgy of violence, with the gang ultimately routed by the incensed townsfolk.

"We Want to Get Loaded"

Griffith's original script focused on the hunger for thrills shared by both the outlaw bikers and the motorcycle cops who hunt them. But Corman wasn't impressed and asked for substantial rewrites. In Griffith's second version, subsequently redrafted by Peter Bogdanovich, the cops sank into the background. Instead, the bikers' alienation and anarchy were given full play.

The Wild Angels (originally titled *All the Fallen Angels*) is a mischievous pageant of excess, which goes all out to exploit the shock value of the outlaw biker. Feeding off the shrill panic that surrounded the Hell's Angels, Corman's picture turns the biker gang into grotesque savages who jar the audience's sensibilities and sneer at their dreary conventions. With a threadbare plot and minimalist script, it's this gleeful emphasis on transgression that is central to *The Wild Angels*. Shots of open kissing between male gang members consciously courted

Wild Angels

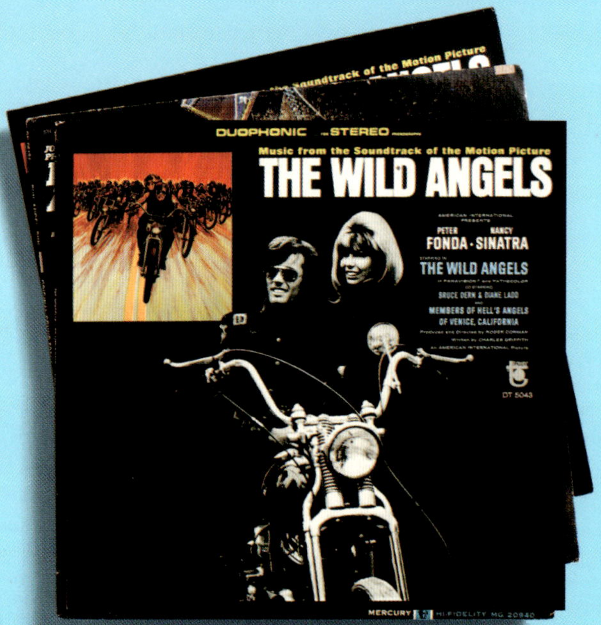

Like the film, publicity for *The Wild Angels* was rough and raucous.

outrage, while everyday norms dissolved in the portrayal of the Angels' wild partying and constant brawling, their sexual violence, and (yes, you guessed it) their invasion of a small-town backwater.

The mayhem culminates in the local church (where else?). At Loser's funeral, the priest is perturbed by the antagonism of the bikers facing him. "Just what is it that you want?," entreats the perplexed minister. The response from Blues, the gang leader, is a salute to disaffected and directionless revolt. "We want to be free" Blues proclaims, "We want to be free to do what we want to do... We want to be free to ride our machines without being hassled by The Man. And we want to get loaded."

"We want to get loaded"—Peter Fonda sticks it to The Man.

ROGER CORMAN

Movie producer and director Roger Corman became famous as a low-budget schlock-meister. His career took off in the early 1950s and (renowned for his fast and furious approach to moviemaking) Corman produced up to seven movies a year until his retirement in 1971. His fastest movie was probably The Little Shop of Horrors *(1960)—shot, by all accounts, in just two days and a night.*

The grim brutalism of The Wild Angels *was typical of Corman's movies. Many of his pictures depicted modern life as arbitrary, meaningless, and terrifying. But there was also a dark humor to Corman's work. Flaunting his low budgets and peppering his movies with elaborate in-jokes for media-savvy audiences, Corman's pictures were often ironic and tongue-in-cheek.* The Wild Angels *was typical, with an over-the-top carnival of transgression that was calculated to tweak the tails of uptight moralists.*

All the Fallen Angels

After Blues' rabble-rousing speech, Loser's remembrance service degenerates into sleazy bedlam. In an extended sequence of drunken revelry (nearly 15 minutes long), a procession of social taboos is brazenly flouted. In the background, the church crucifix is neatly juxtaposed against the swastika flag draped over Loser's coffin as the dirty, unshaven bikers smash the pews, rough up the priest, and embark on a binge of drink, drugs, and sordid sex. To a soundtrack of wild bongos (what else?) a scene of primal savagery ensues. Frenzied dancers gyrate in abandon, bikers stumble about in a drug-addled daze, and a girl has sex with assorted gang members behind the church alter. Hoisted from its coffin, Loser's sagging corpse is propped against a wall, a lighted joint placed comically between its lips, while the bound and beaten priest is dumped in the empty casket. Loser's girlfriend (Diane Ladd), meanwhile, is brutally assaulted, drugged, and then raped (off screen) by two of the dead man's erstwhile comrades.

The bleak nihilism of *The Wild Angels* is underscored by the movie's conclusion. Asked to say something at Loser's graveside, gang leader Blues can only reply "There's nothing to say." Police sirens finally sound in the distance and someone encourages Blues to flee, but all he can do is answer "There's nowhere to go," and begins shoveling dirt into his friend's grave as the movie's credits begin to roll.

Full-Throttle Cinema

***Hell's Angels '69* had a neat bonus—it starred *real* members of the Hell's Angels Oakland chapter.**

Initial responses to *The Wild Angels* were hostile. At a convention of the Theater Owners' Association in New York, previews of the movie provoked disgust. As the picture played, AIP bosses Nicholson and Arkoff nervously watched a steady exodus into the lobby—one cinema owner's wife asking "What kind of people are you to make a picture like this?" Entered into the 1966 Venice Film Festival, the movie got a mixed response, but was reviled in press reviews—*Newsweek* speaking for many when it dismissed the picture as an "ugly piece of trash."

But the bad publicity and outrage was a commercial godsend, effectively guaranteeing the movie's credentials as a scandalously shocking spectacle. As a consequence, *The Wild Angels* became a huge hit—especially with young audiences and the drive-in circuit. AIP struggled to keep up with demand for prints of the movie as it grossed a hefty $5 million in only its opening month.

The movie was one of AIP's biggest ever successes, yet not everyone was impressed. Feeling shortchanged by the studio, and resenting the way they were depicted in the movie, the Hell's Angels (reputedly) threatened to kill Roger Corman and sued AIP for $2 million, claiming the movie had defamed its image. The club, however, was soon placated with a $200,000 settlement, and the Oakland Angels' president, Sonny Barger, even signed up to work on several of AIP's biker sequels.

Devil's Angels

In all, AIP went on to produce about a dozen biker pics. In 1967, while *The Wild Angels* was still screening, the studio quickly completed *Devil's Angels*. Directed by Daniel Haller and starring John Cassavetes, this was (yet) another tale of small-town locals terrorized by evil bikers. As became the custom in biker pics, the sparse script and storyline gave way to an emphasis on visuals, with panoramic shots of cruising motorcycles and a succession of nihilistic tableaus that (mimicking the overblown press stories) grandstanded the gang's lifestyle of beer-swilling, fist fights, and sexual violence.

In *The Glory Stompers* Dennis Hopper leads a band of scooter trash known as the Black Souls. This particular "chopper opera" features white slavery, chain whipping, a love-in, obligatory switchblade fights, and copious beer swilling.

Hell's Angels on Wheels

But it was an AIP competitor, US Films, who came up with the most successful biker movie of 1967—*Hell's Angels on Wheels*. Produced by "B"-movie veteran Joe Solomon, the movie stars Jack Nicholson as introspective loner, Poet, whose spell on the road with outlaw bikers ends in a violent confrontation with the gang leader over the latter's neglected and abused "old lady." The movie portrayed the usual exaggerated feral biker lifestyle, though had the added spice that the Hell's Angels had endorsed the picture—with the entire Oakland chapter appearing in the opening sequence and Sonny Barger credited as technical advisor. Jack Nicholson also appeared in a second biker pic that year, *Rebel Rousers,* but the movie was a turkey and, although it was completed in 1967, it stayed in the can until 1970.

Hell's Angels on Wheels, on the other hand, was a hit. Convinced of the biker pic's bankability, Joe Solomon followed up quickly with *Angels from Hell*. Released in 1968, the movie stars Tom Stern as a Vietnam veteran left bitter and resentful by his wartime experience. In revenge, the disgruntled vet organizes the biggest, meanest, ugliest biker gang in history. Five-hundred strong, the Angels from Hell descend upon (you guessed it…) a small, rural town to exact vengeance.

The following year Solomon produced his third biker epic with *Run Angel, Run*, starring William "Big Bill" Smith in his first biker role. Big Bill plays Angel, a tough biker, who breaks the outlaw code by selling a scoop on his gang to a magazine for $100,000. Angel then takes to the road, pursued by his former buddies who are out for revenge. Shot in 13 days for under $100,000, *Run Angel, Run* proved another big money-spinner, earning around $13 million.

Down and Dirty Cycle Trash

A consummate huckster, US Films' Joe Solomon was renowned in the business of exploitation moviemaking. "Exploitation movie" is a vague term, often used simply to describe movies that critics regard as vulgar or tasteless. But, in the movie industry, the phrase has a more precise meaning. Here, the term "exploitation" denotes low-budget movies containing scenes of sex, violence, or other lurid elements that promoters capitalize on with salacious advertising.

The 1950s and 1960s were the "golden age" of the exploitation movie. As mainstream audiences declined, the major studios pulled back from making low-budget, B-movies, leaving the field open to independent moviemakers, keen to turn a quick buck. A relaxation of censorship also ensured that more taboo subjects became fodder for the exploitation merchants, while the lucrative youth and drive-in markets provided a ready demand for schlock.

As a kind of "tabloid" cinema, exploitation movies were often tied in to current fads or controversies. Hardly surprising, then, that a posse of independent moviemakers (alongside AIP and Joe Solomon) took full advantage of the 1960s biker panic. William Greffe, for instance, leapt into the saddle with *The Wild Rebels* (1967), and Titus Moody (who went on to become a 1970s porn movie pioneer) released *Outlaw Motorcycles* (1967), and *Hell's Chosen Few* (1968), while K. Gordon Murray (a kingpin of the exploitation market) offered *Savages from Hell* (1968).

A particularly visceral example of the down-and-dirty "biker–sploitation" genre is *Wild Riders*. Released in 1971, the movie features two ruthless, amoral bikers who get kicked out of

The archetypal outlaws of classic biker pics were wild, wanton... and usually pretty wicked.

their Florida motorcycle club for being just a bit too mean (having nailed a girl to a tree). Traveling west to California, the pair happen upon two women alone in a mansion. With echoes of the Charles Manson murders (then hugely newsworthy), the bikers invite themselves in for a party of rape, terror, and bloodshed. The two goons are finally brought to heel when the husband of one of the victims returns home. A concert musician, he takes out the thugs with his cello bow (yes, death by cello!).

Hell's Angels '69

Despite the competition from US Films and others, AIP was always leaders of the biker movie pack. Throughout the late 1960s biker pics rolled off the studio's production-line—AIP often recycling props and using the same actors and crew from one movie to the next. In 1967 AIP followed up *Devil's Angels* with *Born Losers* (1967), featuring writer/director/actor Tom Laughlin in the role of half-Indian activist Billy Jack, who stands up against a predictably obnoxious biker gang. In the same year AIP also released *The Glory Stompers*, with Dennis Hopper making his biker movie debut as the unhinged leader of the Black Souls Motorcycle Club. Subsequent years saw further AIP biker fare with *The Savage Seven* (1968), *The Cycle Savages* (1969), and *Hell's Belles* (1969).

In *Hell's Angels '69* (1969), the action switches to Las Vegas. Planning to rob the Caesar's Palace casino, two rich kids dupe a group of Hell's Angels into being a disruptive diversion for their robbery. Less than impressed when they find out, the Angels chase the pair through the deserts of Nevada. What makes the picture especially notable, however, is that the biker gang are played by the Hell's Angels Oakland chapter, with figures such as Sonny Barger and Terry the Tramp playing themselves.

"Iron Horse" Epics

Low budget and jammed with sensation, AIP's biker flicks were true to the studio's tried-and-tested formula. But the 1960s' biker movies also drew on myths central to American nationhood. Most obviously, the biker genre drew a close analogy between the motorcycle outlaw and the frontier gunslinger of the Old West. As Roger Corman later recalled, in making *The Wild Angels*, he "saw the Hell's Angel riding free as a modern-day cowboy. The chopper was his horse. The locales would be the wide-open spaces—the beach, the desert, and the mountains."

In fact, to make its biker movies, AIP effectively re-tooled its Western movie assembly line—with Western production crews and actors reassigned to the production of "iron horse" epics. Moreover, many biker plots were actually lifted from classic Westerns. For example, the narrative of *Chrome and Hot Leather* (1968) is indebted to the *Magnificent Seven* (1960), while *Hell's Belles* (1969) borrows heavily from the James Stewart movie, *Winchester '73* (1950).

The scores of 1960s "chopper operas" were ideal drive-in fodder—they were cheap, simple ...and crammed with sex, violence, and "sickle" action.

In *Run, Angel, Run* William "Big Bill" Smith made his biker movie debut.

Riding Free...

But the archetypal biker movie also had another appeal. The genre's treatment of the "outlaw menace" was always decidedly ambiguous—superficially the pictures may have inveighed heavily against the lawlessness of rogue biker gangs, but the movies also offered audiences the "subversive pleasures" of reckless and irresponsible kicks. It was an attraction that Joe Solomon knew only too well. The exploitation movie maestro once confided:

> ***You take a motorcycle gang. They're putting down cops, smoking grass, laughing at the law. ... They're taking women by force! Carnage! Pillory! (sic.) They've got money, guns. Some guy on a bike grabs a girl and screws her. No consequences, no kids, riding free ... Kids think, God! I've got a mother, a father, I'm living at home. God! If I could have that freedom...***

The same sense of full-throttle freedom also fed into the outlaws' perception of themselves. As 1960s biker-turned-TV evangelist Barry Mayson explains in his autobiography *Fallen Angel: Hell's Angel to Heaven's Saint:*

> ***It was mostly a lot of noise and fun at first, the biking life. All of us had been influenced by the movies or news media accounts of Hell's Angels or other big bike clubs. So they provided the model we tried to act out.***

> ***We'd storm into a bar sometimes and start raising loud hell. We'd start fights anywhere at the drop of an eyebrow, "duke out" bouncers, do wheelies in the middle of town and generally make as much commotion as we could everywhere we went.***

Screened around the world, the 1960s chopper operas also gave the Californian outlaw international exposure, and ensured that his style and attitudes spread worldwide.

WILLIAM "BIG BILL" SMITH. *"Big Bill" Smith, one of Hollywood's best-known character actors, cut his teeth in the glut of biker pics that appeared during the late 1960s and early 1970s. Off-screen Smith has a reputation as an all-round nice guy, but on the screen his dangerous looks made him a natural mean biker.*

After his biker movie debut in Run, Angel, Run *(1969), Smith went on to star in Joe Solomon's* The Losers *(1970) as "Link," a tough biker recruited by the US Army to rescue a CIA agent held in Cambodia. 1970 also saw Smith star as both bad biker "Moon" in* C.C. and Company, *and as misunderstood biker "Tim" in* Angels Die Hard.

Hollywood Man *(1970) found Smith in a semi-autobiographical role—starring as a motorcycle gang leader who sinks all his cash (as well as some of the mob's) into making a biker movie. Big Bill also reprised his "bad biker" role in 1986, starring in* Eye of the Tiger.

"Sickle Sounds"

Raw, fast, and dirty—motorcycles and rock 'n' roll were virtually made for each other. Elvis Presley loved bikes and *Roustabout* (1964), in which he stars as an itinerant, two-wheeled songster, is one of the King's better movie appearances. But rock 'n' roll's real "motorcycle bad boy" was Gene Vincent.

Race With the Devil

Vincent was one of rock's first mythic icons. As a 20-year-old in the US Navy, his left leg was crushed to a pulp in a motorcycle accident, leaving him with a permanent limp and chronic pain for the rest of his life. Gradually recovering, Vincent concentrated on building a music career, fronting his band, Gene Vincent and the Blue Caps. Signed up by Capital (who hoped they'd found a rival to Elvis), Vincent scored his greatest hit, "Be-Bop-a-Lula," in 1955. Followups like "Race With the Devil" and "Bluejean Bop" (both 1956) were liked by the critics, but their lack of commercial success prompted Vincent's eventual relocation to England in 1959.

Taken in hand by British rock 'n' roll impresario, Jack Good, Gene Vincent underwent a total image makeover. In place of his rock-a-billy panel shirts, Vincent adopted the black leather look for which he was to become famous. Assuming the stage persona of a demonic biker, Vincent's performances were rock 'n' roll at its most frenzied and powerful, and earned him regular appearances on Good's TV music show, *Boy Meets Girl*. A tour of Britain in 1960, however, brought tragedy when Vincent and friend Eddie Cochran (who shared the bill on Vincent's shows) were involved in a car accident. Cochran was killed, and Vincent was left traumatized. In subsequent years the singer's recordings were increasingly sporadic and his heavy drinking worsened. Gene Vincent was eventually to die from his alcoholism in 1971—aged only 36.

Elvis Presley loved bikes, but *Roustabout* is the only movie in which The King stars on two wheels.

The early 1960s were the heyday of "motorcycle pop," with a glut of singles and LPs that celebrated the "bad biker" image.

"Sickle Sounds"

Leaders of the Pack

The 1950s and 1960s, however, also saw lighter takes on the motorcycle myth. In 1955, for example, the famous songwriting partnership, Jerry Leiber and Mike Stoller, penned "Black Denim Trousers," a song that provided a hit for vocal trio The Cheers. Covered the same year by clean-cut vocal group The Diamonds, the track gave a chirpy spin to the "bad biker" image:

> ***He wore black denim trousers and motorcycle boots***
> ***And a black leather jacket with an eagle on the back***
> ***He had a hopped-up sickle that took off like a gun***
> ***That fool was the terror of Highway 101***

Girl groups, too, took up the motorcycle theme. In 1962 Phil Spector's prodigies, The Crystals, had a #1 hit with "He's a Rebel"—an anthem to the romance of teen rebellion written by Gene Pitney:

> ***He's a rebel and he'll never ever be any good***
> ***He's a rebel cos he never ever does what he should***

But it was the Shangri-Las, whose adolescent charm was tinged with more than a hint of darkness, who had all the attitude. Topping the charts in 1964, "Leader of the Pack" was the group at its melodramatic finest. Punctuated by motorcycle roars and ending in a crescendo of crashing glass, the song was teen anguish on an operatic scale.

Known as the "Myrmidons of Melodrama," the Shangri-Las topped the charts in 1964 with "Leader of the Pack."

Written by Beach Boy Brian Wilson, "Little Honda" was a big hit for the Hondells in 1964.

Motorcycle Pop

The early 1960s were the heyday of "motorcycle pop"—light, perky tunes that celebrated the fun to be had on the small, easy-to-ride Japanese machines then conquering the American motorcycle market. In 1964 the Risers scored a minor hit with "She's a Bad Motorcycle," but the field was dominated by the Kickstands and the Hondells—both the brainchild of Californian music legend, Gary Usher.

It's difficult to overstate the importance of Usher to the history of West Coast rock. The LA-based writer/producer cowrote several classic Beach Boys' songs with Brian Wilson, and produced albums for the Byrds. The Kickstands and the Hondells were both Usher's pet projects—the "bands" were, in fact, composed of floating line-ups of session musicians overseen by Usher at the mixing desk.

The Kickstands officially recorded one album, *Black Boots and Bikes*, on Capitol Records in 1964. With tracks like "Death Valley Run" and "Mean Streak," it was in a darker vein than the Hondells' releases. The Hondells' biggest hit was 1964's "Little Honda." Written by Brian Wilson, it was a chipper celebration of small Japanese bikes, and scooted up to number nine in the Billboard chart. In the wake of the song's success, a touring version of the Hondells was put together, but ace LA session men such as guitarist Glen Campbell and drummer Hal Blaine always made the group's recordings. The Hondells never had another hit on the scale of "Little Honda," but went on to release two albums and 13 singles before cutting their engine in 1970.

Nashville bassist, Bob Moore also got in on the act. A bass-player extraordinaire, Moore had played with such stars as Elvis Presley, Roy Orbison, and Brenda Lee, but he also cut tracks under his own name. In 1966 he recorded the single "Hell's Angels"—a timely release, though its lighthearted take on the topic saw it quickly fall by the roadside.

Loud, Loose, and Savage

AIP's biker movies heralded the arrival of a more hard-riding approach to motorcycle rock. Roger Corman's approach to moviemaking gave short shrift to Hollywood convention, with his pictures' storylines and structures giving way to spectacle—and a trademark of his 1960s chopper operas was their high-octane sequences of "sickle action" played out to pulsating guitar riffs.

The trend was kickstarted by the soundtrack to *The Wild Angels*. Provided by Davie Allen and the Arrows, the score's dirty, fuzzed-up power chords were ideally suited to the movie—they were loud, loose, and savage. The soundtrack to the movie also powered up the Billboard album chart, while Allen went on to provide an array of scores for AIP movies throughout the late 1960s and early 1970s.

DAVIE ALLAN AND THE ARROWS

Guitar legend Davie Allan is the undisputed king of the biker movie soundtrack. During the late 1960s and early 1970s he carved a niche in the annals of rock with a salvo of classic instrumentals and two dozen movie scores.

Allan formed his band, the Arrows, while in high school at Van Nuys, California. Session work brought interest from movie producers, and in 1965 the band provided the soundtrack for Skaterdater, *an 18-minute short about skateboarding. A big break came the following year when Allan was signed up by AIP to provide the soundtrack for* The Wild Angels. *Allan's distinctive combination of expressive minimalism and "wall of fuzz" guitar playing suited the biker genre perfectly and he went on to provide scores for* Devil's Angels, Born Losers, The Glory Stompers, The Hellcats, *and many more.*

The 1990s found Davie Allan back in the saddle with a fresh volley of biker-inspired albums.

Easy Rider

Easy Rider was the biggest biker movie of 1969. The picture first took shape in 1967 when Peter Fonda and Dennis Hopper were both working on _The Trip_, an AIP "drugsploitation" movie written by Jack Nicholson and directed by Roger Corman. When Corman was reluctant to shoot an acid-trip scene, Hopper and Fonda went into the desert and shot the sequence themselves. Enjoying the experience so much, the pair resolved to continue their moviemaking partnership.

Originally titled *The Loners*, Hopper and Fonda's project was nearly an AIP production. As a writer/producer/director team, Fonda and Hopper had offered the movie to AIP boss Sam Arkoff. But, unsure of Hopper's directorial abilities, Arkoff demanded the right to bring in a new director if production fell behind schedule. Reluctant to risk losing control of their baby, Fonda and Hopper took the project to producers Bert Schneider and Bob Rafelson, who secured big league backing from Columbia Pictures. It proved a shrewd move. Though Fonda and Hopper worked with a budget of less than $400,000, *Easy Rider* was a major box office hit, grossing over $19 million in its first year of release.

Head Out on the Highway...

Acclaimed by the critics, *Easy Rider* is widely regarded as the first bona-fide road movie. The screenplay was written by Hopper and Fonda, along with Terry Southern, an established movie writer. Southern, however, was a late addition, drafted in to give the project an air of gravitas. The picture itself was directed

Riding free—Hopper, Fonda, and Nicholson hit the road.

(somewhat maniacally) by Hopper, and was shot in less than two months—including a week spent in New Orleans shooting the Mardi Gras and graveyard sequences.

The movie begins in the Californian desert, with two freewheelers Wyatt (aka Captain America—played by Peter Fonda) and Billy (Dennis Hopper) making a big cocaine sale to an LA connection (Phil Spector in a cameo role). With the cash concealed in the stars-and-stripes gas tank of Wyatt's gleaming chopper, the pair ride eastward to party at the New Orleans Mardi Gras before "retiring" to Florida. But their motorcycle odyssey through the American Southwest is interrupted when they are hassled by small-town cops and slung into jail for "parading without a permit." The two are befriended by their cellmate, a drunken civil-rights lawyer called George Hanson (Jack Nicholson), who manages to spring them from the slammer and then joins the pair on their road trip. Around the campfire, Wyatt and Billy introduce Hanson to the joys of smoking grass, but things soon turn sour. After the trio are baited by rednecks in a diner, an axe-wielding gang clubs Hanson to death.

Billy and Wyatt continue to New Orleans, pulling in to a brothel that Hanson had recommended. There, they celebrate Mardi Gras and drop LSD in a cemetery with two young prostitutes (played by Karen Black and Toni Basil). Then, still with their cash safely stashed, the two hit the road, Billy laughing triumphantly, "We've done it. We're rich, Wyatt. We did it." But Wyatt's cryptic response is simply, "We blew it."

The freedom of the open highway is short-lived. As Wyatt and Billy ride their bikes down a rural back road, the two are brutally gunned down by shotgun-toting rednecks in a pickup truck.

PETER FONDA

Peter Fonda's early career took shape under the shadow of his famous father, Henry. Initially cast in romantic leads, Fonda found a new direction when he featured as a motorcycle gang leader in Roger Corman's The Wild Angels *(1966).*

As Wyatt/Captain America in Easy Rider, *Fonda embodied the spirit of a nation desperately searching for a sense of itself. It was a theme underlined by the stars and stripes resplendent on the back of the character's jacket, his motorcycle helmet, and his bike's gas tank. Fonda had actually worn the jacket and ridden the bike around LA a week before shooting began on the movie—and was pulled over several times by cops who took exception to his irreverent flaunting of the national flag.*

Easy Rider *earned Fonda an Oscar nomination for Best Screenplay, and contributed to Hollywood's growing interest in young audiences and socially relevant movies.*

Easy Rider

George Hanson (Jack Nicholson) and Wyatt (aka Captain America—played by Peter Fonda) go "looking for America" in *Easy Rider*.

Born to Be Wild

In some respects, *Easy Rider* took the classic themes and conventions of the biker pic to a mainstream arena. Fonda, Hopper, and Nicholson were all biker movie veterans, and *Easy Rider*'s visual style, quick-paced editing, and improvised dialog gave the movie a sense of raw spontaneity akin to that developed in *The Wild Angels* and its sequels. The fierce rock soundtracks of the biker flicks are also echoed in *Easy Rider*'s score of contemporary rock songs by the likes of Jimi Hendrix, The Band, and (especially) Steppenwolf.

Like the biker classics, *Easy Rider* also features many Western allusions. The main characters' names (Billy and Wyatt) are reminiscent of cowboy gunfighters, while Billy's buckskin coat and Stetson are obvious Western touches. The journey across the vastness of the American landscape also invokes the pioneering spirit of the early settlers and the innumerable Western movies that mythologized them.

"We Blew It"

But *Easy Rider* gave a more subtle spin to the biker pic formula. Whereas a relish for reckless kicks was the hallmark of the "leather and chrome" pack, *Easy Rider* is a disconsolate social commentary. With its wistful tag-line "A man went looking for America. And couldn't find it anywhere," together with Wyatt's poignant remark, "We blew it," *Easy Rider* was an allegory for the collapse of American ideals in the face of burning inner-city ghettoes and the trauma of the Vietnam War. As the lawyer Hanson sagely explains to Wyatt and Billy, the American myths of liberty and freedom have become an illusion:

> ***This used to be a helluva good country. There's a lot of talk about freedom and the individual, but no freedom. Show the people a little freedom and they're terrified.***

Footloose and enigmatic, ***Easy Rider*** **was the biker box office hit of 1969.**

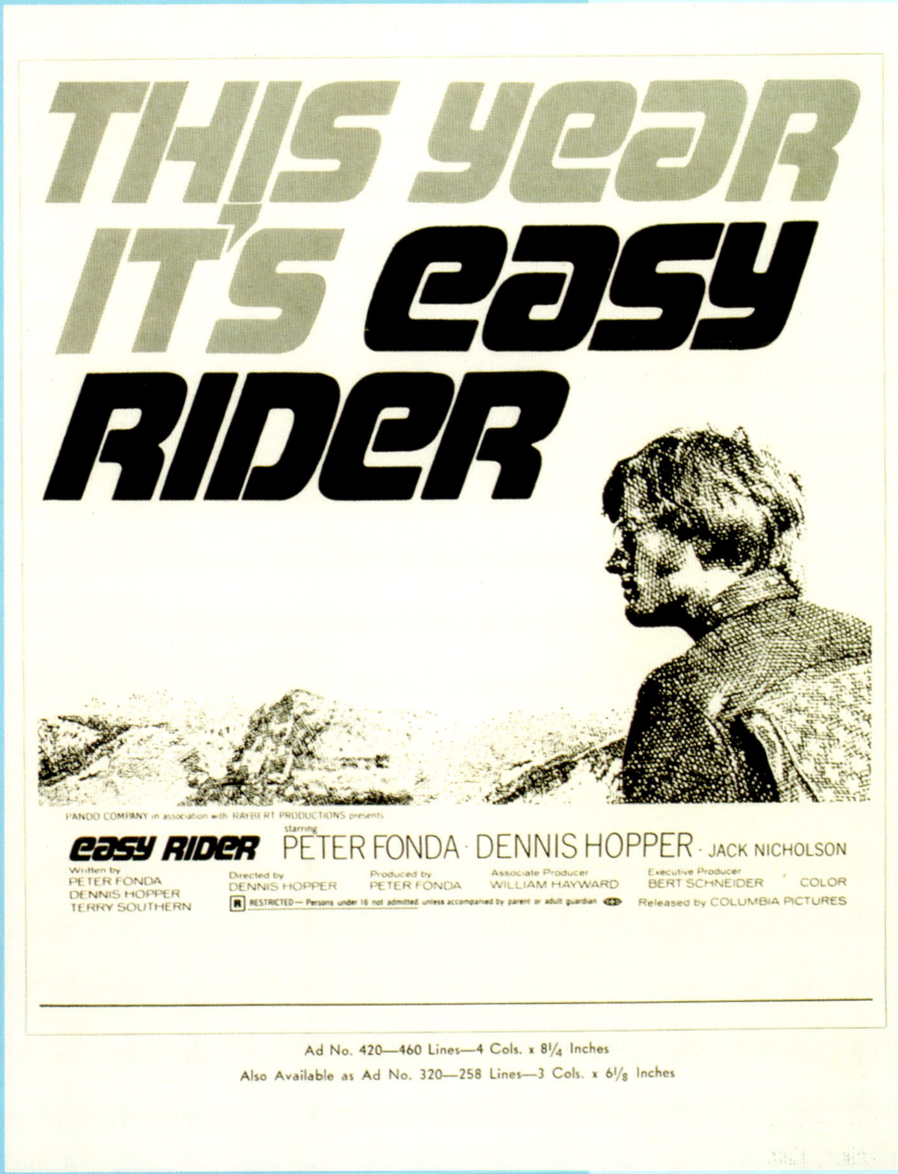

JACK NICHOLSON

Jack Nicholson got his first big break in Easy Rider. *In the role of alcoholic civil-rights lawyer, George Hanson, he earned an Oscar nomination for "Best Supporting Actor" and quickly shot to movie stardom.*

Spotted by Roger Corman in the late 1950s, Nicholson was cast as the lead in Corman's 1958 quickie, The Cry Baby Killer, *before being given the lead in* Hell's Angels on Wheels. *In 1967 he wrote an LSD-hippy drama,* The Trip, *directed by Corman and co-starring Dennis Hopper and Peter Fonda.*

Casting Easy Rider, *Hopper and Fonda initially lined up actor Rip Torn for the part of Hanson. When it transpired that Torn was unavailable, however, Nicholson was offered the role. Making the character his own, Nicholson turned Hanson into an eccentric savant.*

DENNIS HOPPER

Following a lean period in the late 1950s and early 1960s, things picked up for Dennis Hopper when he was cast in the psychedelic movies The Trip *(1967) and* Head *(1968). During the making of these movies he met up with Peter Fonda and Jack Nicholson and began developing the ideas that led to* Easy Rider.

The success of Easy Rider *saw Hopper hailed as a major new moviemaker. But he had been difficult to work with—many people quit the set due to Hopper's drug-induced paranoia and rages.*

After months spent editing Easy Rider, *Hopper finally delivered a 4.5-hour epic. Columbia executives weren't impressed, and pressured Hopper into cuts. Months later the movie was still 3 hours long, so—with Hopper in New Mexico—an editor named Don Cambren was enlisted to trim the movie down to a neater 95 minutes. On return, Hopper was livid, but grudgingly agreed to the changes.*

End of the Road for the Biker Movie?

The box office success of *Easy Rider* brought a new wave of silver screen riders. In 1969 *Wild Wheels* saw bikers battle it out with a dune buggy gang, while Roger Corman returned to the road with another tale of murderous bikers in *Naked Angels* (1969). By this time, however, the genre was flagging and producers increasingly sought different slants and new gimmicks. In *The Dirt Gang* (1972), for example, vicious renegades take to dirt bikes as they terrorize a movie crew in a desert ghost town.

Another take on the genre, in 1969, came with the arrival of the "good guy" biker, in the form of Joe Namath (star quarterback of the New York Jets) taking the lead in *C.C. and Company.* And in *The Losers* (1970), outlaw bikers found themselves on the same side as "The Man," when a motorcycle gang joins forces with the CIA to bust out American POWs from a Cambodian prison camp.

The Wild, The Wicked, and The Weird

Biker/blaxploitation crossovers also appeared. *The Black Angels* (1970), for instance, features a violent rivalry between black and white biker gangs. And the same theme surfaces in the *The Black Six* (1974), except this time the black gang are played by six beefy football stars led by "Mean" Joe Greene of the Pittsburgh Steelers. A more comic inflection came in *Darktown Strutters* (1974), which features a tough gang of cycle-riding soul sisters confronting racist cops to an infectious disco soundtrack. However, even this somewhat surreal offering cannot match the undiluted kitsch of *The Pink Angels* (1971), which features the antics of a group of high-camp bikers, The Cupcakes, on route to a transvestite ball in Los Angeles.

Weird and wild biker/horror hybrids also surfaced. *Werewolves on Wheels* (1971) found biker gang The Devil's Advocates battling satanic, black-robed monks. In Britain, meanwhile, the semi-parodic *Psychomania* (1971) introduced undead bikers. Here, a nihilistic biker (Nicky Henson) decides to make the name of his violent bike gang—"The Living Dead"—more than just a slogan. Learning the secret of eternal life from his mother, a frog-worshipping occultist, the gang leader dives to his death—only to leap out of his grave, still astride his roaring motorcycle. His undead biker pals soon join him, and the gang raise hell throughout the English countryside.

"The Living Dead"—the biker gang risen from the grave in *Psychomania*.

In spite of the new hooks, however, audiences were getting bored with biker movies. The genre spluttered on through the early 1970s, but increasingly ran out of road.

THE DIRT GANG
PAUL CARR · MICHAEL FOREST · BEN ARCHIBEK
71/121
JOE SOLOMON
THE LOSERS
WILLIAM SMITH
BERNIE HAMILTON
as Captain Jackson
ADAM ROARKE
as Duke
In COLOR
HUSTON FEERLESS · ANA KORITA · JOHN GARWOOD · PAUL KOSLO · GENE CORNELIUS
Produced by JOE SOLOMON · Associate Producer VINCENT NAYVE · Directed by JACK STARRETT · Screenplay by ALAN CAILLOU
PRODUCED AND RELEASED BY FANFARE FILM PRODUCTIONS, INC.
THE PINK ANGELS
– IF YOU CAN!
EASTMANCOLOR
71/310
BLAC
ANGEL
OLOR
a portrait of the family.
MERRICK INTERNATIONAL FILM STARRING
LINDA JACKSON · JOHN K
CLANCY SYRKO · BEVERLY GARDNER · JAMES WHITW
STUART FLEMING
DON GORE
71/259
WEREWOLVES ON WHEELS
In COLOR R

BY JAN HUDSON
HELL'S ANGELS
GC214
THE HELLCATS
by ROBERT F. SLATZER
HOLLOWAY
A Fawcett Gold Medal Book
THE BLOOD CIRCUS
Thomas K. Fitzpatrick
231-01971-
The Pack
Brian Black
SOFTCOVER LIBRARY
THE TICKET
Chris Stratton
THEN CAME BRONSON
William Johnston
ROCK!
CHRIS STRATTON
The Bikers
Alex Stuart
CHOPPER Peter Cave
SPEED FREAKS
Peter Cave
ROGUE ANGELS
PETER CAVE
THE OUTLAWS
Alex R. Stuart
Mama
PETER CAVE
RICHARD ALLEN
Alex R. Stuart

03

PULP BIKERS

Outlaws in Print

The World of "Biker-Sploitation"

As well as moviemakers, publishers of pulp fiction were also drawn to the outlaw biker. Breathless press accounts had established in the public's mind the notion that America was infested with biker trash who committed unspeakable acts of sex and violence—and throughout the late 1960s and early 1970s the publishers of an array of pulp magazines and paperback books gleefully cashed in on the panic.

Like the classic biker flicks, the world of the "pulp outlaw" was peculiarly schizophrenic—he was a figure who elicited both abhorrence and admiration. Just like the mainstream media, the pulp industries projected society's fears of lawlessness and random violence onto a grotesque biker caricature. However, these images also had an exciting frisson of the illicit. The pulp bikers' free-and-easy lifestyle was depicted as an endless party of sex, drugs, and adventure. For a workaday Joe, with a job and family responsibilities, it was a heady—and perhaps quite appealing—cocktail of thrills.

A Sideshow of Sleaze

In the back alleys of American publishing, the outlaw biker won a firm following. Downmarket magazine publishers such as Seven Seventy and Press Arts specialized in softcore pornography that tantalized readers with a glimpse of the sleazy, the seamy, and the sensational. Titles like *Banned*, *Barred*, and *Shocker* were a salacious parade of transgression in which the image of the nasty biker found a natural home. During the late 1960s Seven Seventy, in particular, put out several magazines with outlaw bikers as the main theme, including *The Outlaws*, *Sunset Strip Revolt!*, and *Banned: Hells Angels*. Like a carnival sideshow, these magazines' way-out photo-spreads exhibited the biker lifestyle as a startling spectacle of lustful depravity.

"Like Marauding Cavalry"

Men's adventure magazines also took a keen interest in bikers. The origins of the adventure genre go back to titles such as *Argosy*, which began publication in 1882. But during the 1940s and 1950s men's adventure magazines went through a renaissance led by the success of titles such as *True* (launched in 1936), *Stag* (1942), and *Real* (1952). Champions of the hardened tough-guy, these magazines' stock-in-trade was fiction and features dealing with hunting, exploration, crime-fighting, sport, and other "manly" exploits. Along with their taste for adventure, *True* and its ilk also had a penchant for titillation—and they eagerly latched onto bikers as a theme that promised hard-bitten action and lurid prurience.

True was one of the first adventure magazines to spotlight the outlaw biker's libertine lifestyle. Hot on the heels of the Lynch Report's shocking "revelations," *True* initiated its own probing investigation into the "motorcycle menace." Published in August 1965, *True*'s report didn't pull any punches, jolting readers with a histrionic account of two-wheeled savagery:

> ***They call themselves Hell's Angels. They ride, rape, and raid like marauding cavalry—and they boast that no police force can break up their criminal motorcycle fraternity.***

True faced hot competition, its rivals rushing out their own photo features that presented the outlaw biker in all his outrageous glory. In April 1966, for example, *Real* magazine came up with a startling, 10-page Hell's Angels feature produced by Bob Grant, a writer-photographer who had hung out with the club for several weeks. Grant's (often excellent) photos were also the staple ingredient in a one-time-only magazine special. Published in February 1966, *The Real Story Behind The Hell's Angels* purported to be the complete low-down on the club, with "An intimate photo story of their every act—from smoking 'pot' to love making—their women, their bikes, their drunken orgies, their kids." Once again, the narrative was sufficiently ambiguous to portray a biker image that could be seen as glamorous or just plain intimidating:

> ***Suddenly from the distance there was an angry buzz like a cloud of a million angry wasps approaching. Two hundred motorcycles zoomed into town, bearing their wild-eyed jean-clad burdens. A few girls clung possessively to the backs of the riders like love-starved molls. The Hells Angels were on a run!***

Pulp magazine publishers quickly latched on to outlaw bikers as a theme that combined gritty action and outrage.

The Filth and the Fury

By the mid-1960s the original wave of men's adventure magazines was looking tired and dated. Instead, a fresh generation of titles came to the fore. As censorship controls relaxed, magazines such as *Man's Daring*, *Man's Story*, *Men Today*, and *New Man* broke new ground in blood 'n' guts sensationalism, offering their readers escape into a brash world of bravado and adventure.

The new "true adventure" pulp mags were a paean to the grittiest action and the tackiest sex. Purportedly "true" tales of wartime heroics ("There Were Only 12 of Us—Against 5,000 Nazi Beasts!") rubbed up against grainy photo spreads of "cheesecake" models, together with a stream of steamy exposés ("Exposed: The Truth About Love-Hungry Divorcees"). It was a world of wanton sex and violence—and images of filthy, deviant cycle trash fitted in perfectly.

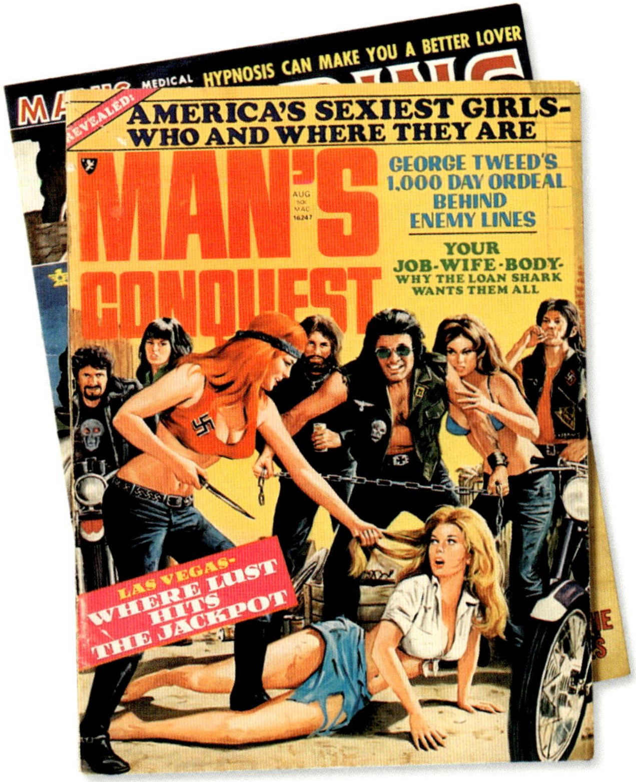

Misogyny, Machismo, and Motorcycles

By now, outlaw bikers were America's foremost bogeymen—so they were a natural choice as malevolent villains in the adventure pulps. The pulps' garish covers and astonishing stories hungrily took up all the news media's "marauding biker" clichés and magnified them a hundredfold. Myriad armies of bikers were depicted visiting a plague of rape and pillage on small-town USA.

The adventure pulps were bellicose in their misogyny. Women were brutalized with mind-numbing regularity in the pulps' stories and outré artwork. And, true to form, the pulp bikers offered women no mercy. A typical tale of torture and torment was 1969's "The Rapers Came to Town," published in *Man's Combat*. "The sleepy little New England village woke up in a hurry when the Huns rode their motorcycles into town!," runs the terse prose, "Attila, their leader, led the wildest mass rape in local history!"

But the evil rarely went unpunished. Invariably, a chisel-jawed hero would arrive seeking revenge and justice (but mainly revenge). A regular theme—that also recurred through the era's biker pics and cheap paperbacks—was that of the Vietnam veteran returning home to find his family ravaged by a loathsome bike gang. "The Cycle Ravishers," a *Man's World* short story of 1969, was standard fare. Bill Sawyer, a marine on leave from 'Nam, returns to his rural Californian hometown. Ambushed by grimy bikers, he is beaten unconscious and his fiancée is abducted. But Sawyer is resourceful. Joined by his four marine buddies, the hero tracks down the bikers, rescues his girl, and punches the gang insensible:

> ***"The Cong were tougher than you punks," Bill said, swinging his fist at one of them.***

High-Octane Thrills

In their yarns of biker mayhem, the adventure pulps always professed righteous anger. Ostensibly, the magazines were appalled by the violence and anarchy they described. But, beneath the veneer of moral indignation, there always lurked a measure of relish for the

The pulps' gaudy covers and far-fetched stories embellished all the media clichés about "marauding biker gangs."

libertine freedom symbolized by the outlaw. For instance, in 1967 *Men* magazine inveighed against the "Rampaging Outlaw Angels and Their Love-Blast 'Mamas'," but the pulp was awe-struck by the bikers' aura of power and effortless cool:

> ***On the back of their stripped-down, souped-up bikes, exhaust pipes smoking, these Swastika-worshiping, chain-wielding "One Percenters"—rebels against the other 99% of the population—and their long-haired, hot bodied females are ready to wreck a bar or work over a "square" at the kick of a starter.***

With women, drugs, and high-octane thrills just there for his taking, the image of the pulp biker offered readers a tantalizing vision of freedom and excitement.

But there was only so much mileage in the formula and, by the end of the 1960s, the pulps' search for new spins and gimmicks produced some ludicrous scenarios. Especially farcical were the "speedboat" bikers that appeared in *Male* magazine's 1968 story, "I Ride With the Outboard Ravagers":

> ***Wet and wild, these hot-throttle water bandits violated every rule of the road, every code of behavior—and every young woman whose bikini-clad body would dress up their 200 HP terror hulls.***

Paperback Bikers

Paperback publishers also eagerly exploited the saleability of outlaw bikers. Ripping their plots straight from the news headlines, pulp novels mesmerized their readers with tales of high-speed kicks in life's fast lane. Originally published in 1958, Bud Clifton's *Road Kids* was typical—with its captivating cover art and a gripping tag-line that ran: "Thrill-crazy kids living for kicks, for girls, and for speed..."

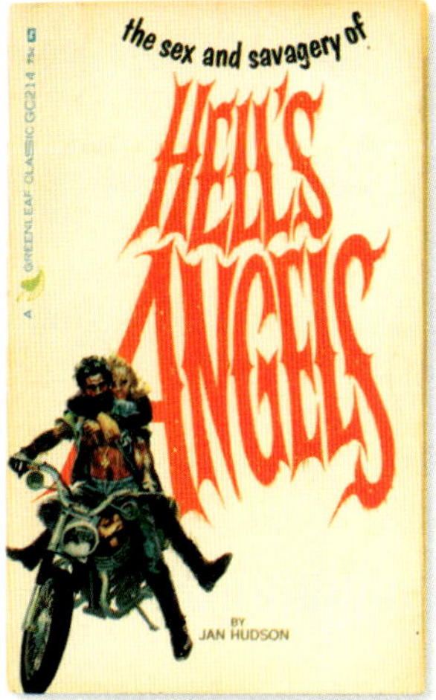

A Saga of Sex and Savagery

The mid-1960s furor surrounding the Hell's Angels spawned a welter of seedy paperbacks that gave full play to the media's sleazy tales of outlaw hell-raising. In 1966 the market was cornered by Jan Hudson, whose *The Sex and Savagery of Hell's Angels* gave the Lynch Report a run for its money with its salacious account of sex, drugs, rape, murder ... and more sex. Hudson wrote with the air of a clued-up "insider" who knew the scene. As a preface to his book, a glossary of "biker" terms showed readers how they too could "talk like a Hell's Angel for fun and profit," while promotional blurbs touted Hudson as someone who "knows the Angels. He's mixed with them and heard their views on life, sex, and violence." The reality, however, was more prosaic. "Jan Hudson" was actually one of many pen names used by George H. Smith, a hack writer who probably culled most of his information from press clippings.

Hard, Fast, and Violent

A flood of rough-and-ready novels matched the juicy sensationalism of Hudson's "factual" exposé. In 1967, for example, *The Pack* featured The Psychos, a motorcycle gang who "rode their midnight-gleaming bikes and loved their women hard and fast, and with unquenchable lust." In *The Blood Circus* (1968), it was The Beasts, a gang who "have a grudge against the whole square world. ... And a taste for unholy terror." *The Bikers* (1970), meanwhile, proclaimed itself as a tale "as violent, sexual, and bizarre as the speed-etched world of men on wheels!"

"The Sweet Ride"

But, occasionally, more sympathetic portrayals also surfaced. In his 1967 novel *The Sweet Ride*, for instance, William Murray offered a relatively unaffected portrait of a motorcycle gang. As a journalist, Murray had earlier profiled the Hell's Angels in an article for the *Saturday Evening Post*. More level-headed than most, his report had dealt with the Angels even-handedly, highlighting the degree of hyperbole in the press panic that surrounded the outlaws. Murray's novel (turned into a motion picture the following year) was equally fair-minded. A murder mystery based around California's surf scene, the book's plot includes an outlaw motorcycle gang—"The 69ers"—that a prejudiced world wrongly accuses of rape and murder.

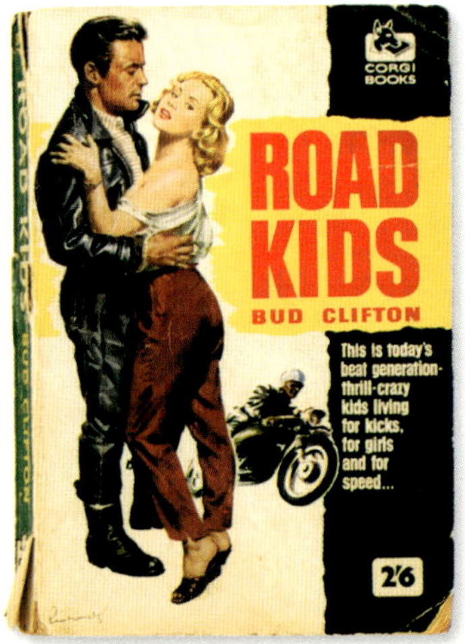

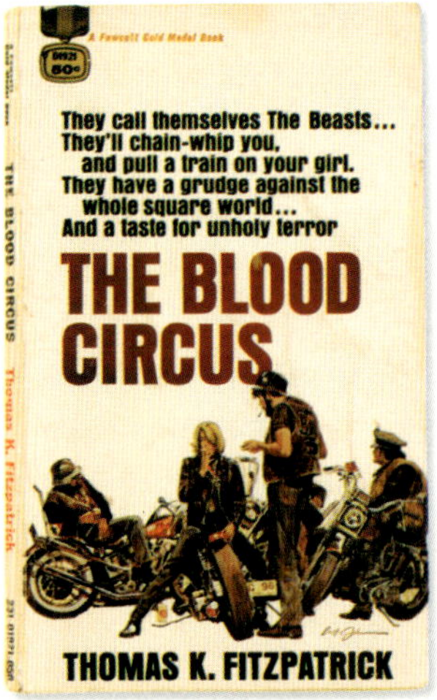

With plots ripped from the news headlines, pulp biker novels enticed readers with their racy artwork and gripping tag-lines.

Legends of Liberty and Brotherhood

Throughout the 1970s, 1980s, and 1990s, the mean biker remained a stock stereotype in the world of pulp fiction. In D.A. Hodgman's detective novel *Deathride* (1992), for example, police agents infiltrate a gang of outlaw bikers who are caricatured as an ugly bunch of gun-running drug-dealers. Biker lowlife are also the villains in Dan Killerman's *Hellrider* series. Killerman's first novel sees Vietnam vet Jesse Heller become a ruthless vigilante. Astride a big, black Harley, and totting a Smith & Wesson .12 gauge, he dishes out slice-and-dice justice to the "Satan's Avenger's"—malevolent bikers who raped and slaughtered his family. Yet the sequel, *Blood Run*, is more ambivalent. This time Heller is on the run from vengeful bikers who, themselves, are vets who discovered brotherhood in Vietnam but were then treated with contempt on returning to "the World."

A more positive treatment of bikers also features in Robert Baron's *Storm Rider* sci-fi series—*Storm Rider* (1992), *River of Fire* (1993), and *Lord of the Plains* (1993). Baron depicts a post-apocalyptic future in which the United States has become a collection of Orwellian city-states whose subjects are comprehensively regulated and controlled. The last vestige of American freedom and individuality lies out on the Plains, where nomadic biker tribes—the High Free Folk—fashion a lifestyle of liberty rooted in a synthesis of biker tradition, Native American folklore, and the Declaration of Independence.

Brotherhood of Outlaws

The *Storm Rider* trilogy is an escapist fantasy. More firmly grounded in reality was *Brotherhood of Outlaws*, published in 1979. A robust tale of biker outlaws on the run from LA cops, the novel's gritty authenticity was guaranteed by the credentials of its author—Bob Bitchin. A former bodyguard for motorcycle stunt legend Evel Knievel, Bitchin (real name Robert Lipkin) was a huge, tattooed biker who created the hard-core motorcycle 'zines *Biker* and *Biker Lifestyle*. *Brotherhood of Outlaws* celebrated the fraternal belonging central to the biker lifestyle, but the novel was also a polemic against helmet laws. A vehement opponent of the legislation, Bitchin was the national director of the US Helmet Law Protest in 1979, a campaign that brought thousands of bikers to the streets in protest against laws that made the wearing of helmets compulsory.

The Best of British "Bovva"

Like their American cousins, British pulp publishers eagerly capitalized on the kudos of outlaw bikers. In Britain, youth subcultures were headline news during the late 1960s and early 1970s. Newspapers plied prurient tales of skinhead gangs, bikers, and soccer hooligans who roamed the streets committing sadistic acts of "bovva" and "agro" ("bother" and "aggravation"—1970s slang for delinquent violence). The stories were grist to the mill for pulp publishers, and a stream of paperback fiction took full advantage of the shock value of these sensational subcultures.

Barbarians on Wheels

New English Library (NEL) led the way in British "bovva" books, becoming especially infamous for its "Skinhead" novels, written by Richard Allen. But NEL was also quick to jump on the biker bandwagon, by releasing UK editions of several US biker novels, along with Jan Hudson's "factual" Hell's Angels' histories.

NEL also produced homegrown biker fare. Published in 1971, NEL's glossy, full-color Hell's Angels magazines boasted they were "The First Pictorial Guide to the Motorcycle Outlaws of the Seventies." The photos of US bikers were stunning, but the text was a rehash of the usual clichés, while the interview with "Zombie"—billed as an authentic "American Angel"—was laughable, and was actually concocted by the NEL office staff. In 1977 NEL repeated its magazine formula on a bigger scale with *Barbarians on Wheels*, a sumptuous coffee-table tome. Again, however, the great visuals were let down by the hackneyed, cliché-ridden prose. NEL's biker fiction, on the other hand, was often quite inspired.

Rogue Angels

Throughout the early 1970s NEL churned out a legion of biker novels. True to NEL's "youth-sploitation" formula, they were chock-full of scenes of gritty violence and sordid sex. And the covers were always "right there," with eye-catching, dramatic pictures and blood-and-thunder tag-lines—"Lock your doors, the Angels are running again!"

One of NEL's main "biker" authors was Peter Cave. Formerly a writer for soft-core pornography magazines, Cave was recruited by NEL to write *Chopper*, the knuckle-bruising saga of "England's King of the Angels":

> ***Action was where Chopper Harris belonged. Like a moth to a naked electric light bulb, or a junkie to his nearest "fix," a Hell's Angel always gravitated toward the nearest possible fight, or the quickest possible sex, or to somewhere an Angel could do something to prove himself one better than an ordinary pig citizen.***

Hitting the bookshops in 1971, *Chopper* was a commercial coup, selling worldwide and going into seven reprints. Only slightly ruffled by death threats from angry bikers (who felt the author was a cynical opportunist), Cave produced a succession of follow-ups, including *Mama* (1972), *Rogue Angels* (1973), and *Speed Freaks* (1973).

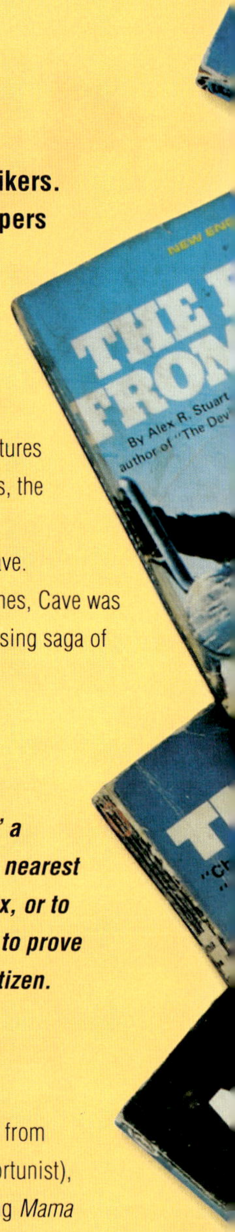

The covers of NEL's biker paperbacks were always eye-catching and dramatic.

NEW ENGLISH LIBRARY AND RICHARD ALLEN

At the beginning of the 1970s New English Library (NEL) was an ailing British publishing firm. But a spectacular change in fortune came when the company shifted direction. Aiming for the youth market, NEL let loose a string of street-tough novels inspired by Britain's headline-grabbing hooligan subcultures.

Skinhead **(1970), by Richard Allen, was the first in the series. Allen's vicious antihero was an East End skinhead—Joe Hawkins—and the novel chronicles his lifestyle of racism, misogyny, and casual violence. The book proved a hit, selling over a million copies and spawning a slew of brutal sequels. After his six "Skinhead" books, Allen wrote more NEL novels that were a veritable demonology of delinquent youth, including** ***Boot Boys*** **(1971),** ***Terrace Terrors*** **(1974), and** ***Punk Rock*** **(1977). Readers assumed that Allen was one of their own, a young thug stalking Britain's soccer grounds. But, in reality, Allen was one of many pen-names used by James Moffat—a middle-aged hack who lived quietly in a seafront cottage.**

Like Cave's greasy creations, Alex Stuart's NEL bikers were cast as a lawless, bone-breaking breed in books such as *The Outlaws* (1972), *The Run* (1972), and *The Bike From Hell* (1973). Mick Norman, on the other hand, offered a more enigmatic version of the genre in his "Angels From Hell" quartet. In tightly written chapters that were intercut with imaginary news reports, press cuttings, and sociological extracts, Norman (a pseudonym for Laurence James, an NEL staff editor) presented a dystopic vision of England in a not-too-distant future. As an oppressive police state grinds down a spineless population, freedom's only hope lies with the Last Heroes—an underground biker gang led by Gerry "Wolf" Vinson:

> ***Screaming like demons from the seventh circle of hell, the surviving Angels, led by Gerry, hurtled out of the quarry into the confused ranks of the police. Any attempt to stop them was futile...***

Beatnik Bikers

In NBC's 1969 TV series, *Then Came Bronson*, Michael Parks starred as a two-wheeled drifter who always leant a helping hand to damsels in distress.

Biker literature wasn't all sex and violence. In the wake of *Easy Rider*, the biker became a totem for free-spirited nonconformity. Published in 1974, *Zen and the Art of Motorcycle Maintenance* turned the motorcycle into a metaphor for spiritual enlightenment. Robert M. Pirsig's chronicle of a ride across Montana was an allegory for man's quest for truth and inner peace.

Then Came Bronson

On TV the biker also got a romantic treatment. In 1969 the NBC series, *Then Came Bronson*, featured Michael Parks (a ringer for James Dean) as Jim Bronson, a beat reporter who quits the rat race and takes to the road astride a mildly customized Harley Sportster. With a mysterious "Eye of God" logo stenciled on his bike's gas tank, Bronson was an enigmatic character and spent the series wandering nomadically along California's coastal highway, running into a variety of adventures against the beautiful backdrop of Big Sur redwood country.

Criticism that *Then Came Bronson* was simply a pallid imitation of *Easy Rider* was a little unfair. The series was in development before the movie's release, and had an appeal all of its own. But, while the Wednesday night episodes won a big following, the network canceled the series after just one season. Nevertheless, to this day, *Then Came Bronson* remains a cult favorite among many Harley riders.

The TV series *Then Came Bronson* also spawned a series of novels.

The Star-Spangled Stunt King

Robert "Evel" Knievel became history's biggest-ever biker star after a series of high-flying (and bone-breaking) motorcycle stunts. Evel Knievel's death-defying career began in 1965. In a touring show called "Evel Knievel's Motorcycle Daredevils" he performed a series of feats, including riding through walls of fire, jumping over mountain lions, and being towed at 200 mph (320 km/h) per hour behind a dragster.

During the 1970s Evel Knievel was famous for his daring and (sometimes barely) death-defying motorcycle stunts.

Going solo in 1966, Knievel's motorcycle jumps became ever longer and more dangerous—climaxing, in 1968, with a 151-foot (45-meter) jump across the fountains of Caesar's Palace in Las Vegas. Knievel cleared the fountains, but his landing was disastrous and the crash left him in a coma for nearly a month. But, as always, the stunt king bounced back. In 1971 he jumped 13 cars in a show that drew a record crowd to the Houston Astrodome. Two years later, Knievel thrilled a crowd of 35,000 when he took off from a towering jump ramp to clear 50 cars stacked in the center of the Los Angeles Coliseum.

The Do-Or-Die Daredevil

With a keen gift for showmanship, in 1974 Knievel hit on the idea of jumping the Grand Canyon. The authorities vetoed his plan, so instead Knievel opted for a do-or-die jump across the Snake River Canyon. Hitting the ignition of his X-2 Skycycle (effectively a rocket on wheels), Knievel soared skyward and cleared the quarter-mile-wide Canyon—but, when the parachutes deployed prematurely, the wind pushed him back into the chasm.

In 1975 another stunt saw Knievel jump 13 double-decker buses at London's Wembley Stadium. After crashing spectacularly on landing, Knievel struggled (despite a broken pelvis) to a microphone and announced to the stunned audience that he would never jump again.

Viva Knievel!

Yet, within six months, a defiant Knievel had returned—jumping 14 Greyhound buses at King's Island in Ohio. But in 1976 disaster struck again. Crashing on his attempt to clear a tank full of man-eating sharks in Chicago, Knievel sustained a concussion and two broken arms. This time the daredevil superstar retired from major performances and allowed his son, Robbie, to assume the mantle of the star-spangled stunt king.

Evel Knievel's daredevil feats made him an American icon. Two movies were based on his exploits—*Evel Knievel* (1971) starring George Hamilton and *Viva Knievel!* (1977) starring Evel as himself—while Evel Knievel toys clocked up sales worth more than $300 million.

Ed "Big Daddy" Roth

Ed "Big Daddy" Roth was a maverick legend of southern California pop culture throughout the 1950s and 1960s. More than anyone else, the artist-turned-custom-car-king was responsible for shaping the esthetics of hot-rod design and the culture that surrounded it. Roth was also a trailblazer of outlaw style. Decades before anyone talked about "subcultures," Roth was familiar with them all first hand—*Time* magazine even dubbed him "the supply sergeant to the Hell's Angels."

The Custom Car Wizard

Ed Roth grew up in southern California during the 1940s, soaking up the fumes of the local custom scene. A hot-rod craftsman par excellence, Roth spent days in his garage chopping, channeling, and spraypainting custom cars that became more than vehicles—they were dazzling phantasmagoria on wheels. Roth's hot-rod creations such as the Outlaw, the Mysterion, and the Beatnik Bandit were a hit on the national show-circuit, and he became the uncrowned king of the 1960s custom-car craze.

The cartoon character "Rat Fink," was also a winner. Originally designed by Roth during the 1950s, Rat Fink was the underground's answer to Mickey Mouse. With his sinister glare, razor-sharp teeth, and bulging, bloodshot eyes the "anti-Mouse" was a ubiquitous presence on T-shirts, posters, and car decals throughout the 1960s.

After Roth was signed up by the Revell toy company in 1962, millions of model kit versions of his cars rolled into the toyshops. And it was a Revell publicity man who came up with Roth's "Big Daddy" nickname, first used as a catchy moniker adorning the colorful kit boxes.

Chopper Mania

Revell's enthusiasm for Roth fell away when he began hanging out with the Hell's Angels. As his interest in customizing motorcycles grew, Roth got to know people on the outlaw bike scene and he came to admire their fiery, rebellious attitude. As he explained to the *Saturday Evening Post* in 1965 "They're the Wild Bill Hickocks, the Billy the Kids—they're the last American heroes we have, man."

Revell, however, wasn't impressed and cancelled Roth's contract in 1967. But "Big Daddy" had no regrets. Going into the publishing business, Roth launched his own biker magazine—*Choppers*. He also built a succession of motorcycle-based customs. Exhibited throughout the US, Roth's outrageously styled machines won an enthusiastic following and firmly established custom bikes as an art form on a par with their four-wheeled counterparts.

The King of Custom

Choppers magazine was to become a major influence on the US custom-bike scene. The magazine not only encouraged the uninitiated to build their own custom bikes, it also provided a medium through which the developing custom-bike industry could swap ideas and sell its wares. *Choppers* was also a boon to Roth's own mail-order business—which did a roaring trade in booklets, T-shirts, posters, decals, and a wealth of other biker-related paraphernalia.

Produced by Ed "Big Daddy" Roth, *Choppers* magazine ran from 1967 to 1969 and was the first magazine to chronicle the world of custom motorcycles.

DAVID MANN

David Mann is widely considered the biker world's artist-in-residence. Born in Kansas City in 1940, his career as an artist kicked off in 1963 when his first painting, "Hollywood Run," scooped a trophy in Kansas City's Custom Car show.

Coming to the attention of Ed Roth, Mann's painting was quickly bought by the Californian custom-car guru who reproduced it as a stunning poster. At Roth's suggestion, Mann moved to the West Coast and met the biker community. Mann went on to paint 14 works for Roth, all capturing the biker's lifestyle, camaraderie, and robust freedom.

*Honing his art, Mann was signed up in 1971 by a new bike magazine—*Easyriders*—and over successive decades his centerfold artwork chronicled the fun, freedom, adventures, and adversities of life as a biker.*

04

THE CYCLE-DELIC YEARS

Bikers and the Counterculture

Sympathy for the Devil

Headline news during the 1960s, outlaw bikers were demonized by the press and hunted by the police. But not everyone was out to get them. Many members of the burgeoning counterculture were beguiled by the outlaws' infamy. For the hippy avant-garde, the biker seemed like a romantic "Noble Savage," and he was eulogized as the personification of raw, spiritual freedom.

A Gathering of the Tribes

The once genteel San Francisco neighborhood of Haight-Ashbury was the hub of the 1960s psychedelic scene; and visiting Hell's Angels became a Haight fixture—their choppers a familiar sight parked outside bars and coffee shops as the riders scored easy drugs and easier women. One Angel, known as "Chocolate George" (for his love of chocolate milk) was a special Haight favorite. After George was killed in a traffic accident, crowds of hippies showed up to his funeral and the *Berkeley Barb*—a local underground newspaper—ran a commemorative drawing of the fallen Angel crowned with a golden halo.

The Hell's Angels even acted as an unofficial police force at flower power "happenings." One of the first was the momentous "Human Be-In" held in Golden Gate Park in January 1967. Billed as a "Gathering of the Tribes," the event was a showcase for beaded hipsterism, with a crowd of 30,000 turning out to see acid rock stalwarts Quicksilver Messenger Service, Jefferson Airplane, and The Grateful Dead strut their stuff. Orations from countercultural gurus such as Timothy Leary also set the tone for the occasion, Leary making his famous entreaty—"Turn On, Tune In, Drop Out." The world seemed groovy and the Hell's Angels, deputized as leather-jacketed stewards, spent the day rounding up lost kids and overseeing the mammoth PA system.

There was a continuing rapport between the Hell's Angels and the Bay Area counterculture. The Grateful Dead, in particular, had a longstanding relationship with the Angels and regularly used them for concert security. To raise funds, the Angels also staged their own events, with shows regularly headlined by West Coast rockers such as Big Brother and the Holding Company.

In the late 1960s the Hell's Angels acted as an unofficial police force at many rock concerts and psychedelic "happenings."

An Unholy Alliance

Psychedelic emissaries Ken Kesey and Allen Ginsberg were both drawn to the image of the biker as a wild renegade. Author of the cult novel *One Flew Over The Cuckoo's Nest* (1962), Kesey had experimented with hallucinogenic drugs during the early 1960s and in 1964 formed the Merry Pranksters—a motley band of hippies who toured the country in a colorfully painted schoolbus. Later, Kesey moved to a mountain ranch in La Honda, overlooking San Francisco, and began throwing parties called "Acid Tests" where revelers imbibed liberal measures of LSD to push back the boundaries of experience. In 1965 Kesey ran into fellow author Hunter S. Thompson, who was then riding with the Frisco Hell's Angels to research his book, *Hell's Angels: The Strange and Terrible Saga of the Outlaw Motorcycle Gangs* (1967). Intrigued by the Angels' anti-establishment reputation, Kesey invited Thompson and his Angel associates up for a weekend at his La Honda spread.

As Tom Wolfe recounted in his countercultural epic, *The Electric Kool-Aid Acid Test* (1968), a column of chopper-riding Hell's Angels duly rolled up to Kesey's ranch on August 7th, 1965. The party raged for nearly three days as outlaws and hippies grooved under strobe lights and blissed-out on the Pranksters' acid. A fellow guest was poet Allen Ginsberg, a charter member of the 1950s beat movement, who galvanized 1960s counterculture through his promotion of eastern spirituality and druggy mysticism. Infatuated with the bikers' edgy aura, Ginsberg quickly drafted a poem in honor of the event—"First Party at Ken Kesey's With Hell's Angels."

To raise funds, the Hell's Angels often staged their own parties and rock concerts.

Strange Brew—The Drugs Connection

The Angels were to become regular houseguests at La Honda, blowing their minds on the reservoir of easy booze, drugs, and women. As the club established a shaky rapport with the psychedelic vanguard, some bikers also became big league LSD dealers. According to some accounts, the Hell's Angels first got involved in the drugs racket in 1964. Facing heavy legal bills during the Monterey rape case (the story goes), some Angels turned to illicit drug deals as a way of generating cash.

Certainly, there was a ready market for drugs in the Bay Area during the 1960s. The Haight-Ashbury hippies devoured colossal amounts of hallucinogens, while further demand came from the hip students at the University of California in Berkeley. LSD was the cool trip of the moment and Augustus Owsley Stanley III was widely regarded as the "Acid King." Usually known simply as Owsley or "Bear," the former chemistry student became legendary for producing potent LSD tabs, and his lab regularly supplied Ken Kesey and the Pranksters. According to George Wethern, former vice president of the Oakland Hell's Angels, Owsley also supplied many bikers. In his autobiography, Wethern claims that the late 1960s saw outlaws increasingly involved in the criminal drugs scene, a number emerging as big-time players. For Owsley, however, the end came in late 1967 when he was sentenced to three years in prison for possession of several thousand tabs of acid.

London Calling

On the other side of the Atlantic, the outlaw biker was also fêted by the countercultural aristocracy. In 1968 several Californian Hell's Angels visited London after hearing reports that unofficial

Like many underground magazines of the period, *Oz* had a soft spot for the Hell's Angels.

During the late 1960s the Hell's Angels were the darlings of the underground press—and were regularly championed in Britain's *Oz* magazine.

"Angels" colors had been spotted in Britain. The visit was arranged by a friend of the club, the rock musician and political activist Mick Farren. Writing in the underground newspaper, the *International Times*, Farren waxed lyrical over the Angels:

> ***The Hell's Angels Motorcycle Club is the American Dream. They are self-proclaimed outlaws, their creed is one of freedom, of pride, of male domination. They drink, they brawl, they act like the Lords of Creation.***

Farren argued for "a mutual exchange and integration of life styles" between the bikers and the counterculture and suggested that, through the dialog, the Angels might find "a deeper philosophy and greater fund of information on which to base their actions." The visiting Angels probably didn't spend too long hunting for "a deeper philosophy," but during their stay they did manage to hook up with some British bikers led by Peter Welsh. Known as "Buttons," Welsh and his associates were granted club charters in 1969 and two chapters of the Hell's Angels were established in the capital—one in South London, the other in East London.

The fledgling British Angels got an early taste of the limelight in July 1969 when they were hired as security for the Rolling Stones' free concert in Hyde Park. It was always planned as a major event, but when band member Brian Jones died two days beforehand, it turned into a huge memorial attracting a crowd of 500,000. The Angels were hired by the Stones' stage manager, Sam Cutler, possibly as a stunt for the Granada TV crew who were filming the show. If so, the bikers didn't disappoint. Resplendent in leather and sawn-off denim, they spent the afternoon guzzling beer and snarling belligerently at anyone who strayed too close.

The British underground press also loved motorcycle outlaws. Countercultural bibles *International Times* (*IT*) and *Oz* both sought "street credibility" by covering the biker scene. *IT* featured a biker cartoon strip called "The Losers," while *Oz* gave over the cover of its April 1969 edition to the Hell's Angels and boasted excerpts from the autobiography of Angels leader, "Freewheelin" Frank Reynolds.

But it was always a fragile alliance. Pacifist, middle-class bohemians and tough, blue-collar bikers made unlikely confederates. In courting the allegiance of the outlaws, the hippies were riding a wild tiger—and the relationship splintered when the Age of Aquarius drew to its ugly conclusion.

HUNTER S. THOMPSON

Hunter S. Thompson established his reputation as a literary gunslinger through his first-hand account of the outlaw biker lifestyle. In 1965 he was dispatched by The Nation *magazine to write a story on the Hell's Angels. The Angels were pleased with* The Nation *article and Thompson soon found himself treated as the outlaws' unofficial ambassador.*

He ultimately spent nearly a year riding with the San Francisco chapter to research his 1966 bestseller, Hell's Angels: The Strange and Terrible Saga of the Outlaw Motorcycle Gang.

But Thompson's relationship with the Hell's Angels ultimately soured. Club members began to resent the author for trading on their notoriety and meted out a severe beating. Despite his mauling, Thompson's book was a runaway success, and with continuous reprints, clocked up paperback sales of more than two million.

"Showing some class" was all about attitude and the exhibition of effortless cool.

Showing Some Class

"Showing some class" was a familiar maxim among outlaw bikers of the 1960s. It denoted the exhibition of style, flair, and panache through acts of daring, defiance, or out-and-out deviance—deeds that would shock the mainstream "citizens" (as outlaws dubbed the general public).

Mean Looks

The biker's philosophy of insolent nonconformity was manifest in his appearance. A biker's long hair, full beard and brawny muscle connoted the ferocious "mountain man," while his grubby Levi's, expansive tattoos, and cold, mean look guaranteed that little old ladies would scuttle across the street to avoid him.

A biker expresses his individuality through his motorcycle—how it's built, how it's painted, how it's ridden. But his jacket is also important, sometimes conveying his personality, beliefs, and experiences through a montage of patches, badges, and pins.

Often worn over a leather motorcycle jacket, the "cut-off" (a denim jacket with the sleeves hacked away) became a stock item in the biker's wardrobe. Emblazoned across its back were the outlaw's colors—the sacrosanct club emblem, surrounded by a top rocker (the club's name) and a bottom rocker (the member's state or province). Worn alone, the bottom rocker identified the biker as a club "prospect" or "striker." A "Front Flash" patch was also sometimes worn over the front left pocket, indicating the member's specific city chapter. "Front patch" clubs also emerged. Less committed to the red-blooded outlaw lifestyle, these clubs distinguished themselves from the "back patch" bad boys by opting for smaller insignia on the front of their jackets.

JACKET PATCHES AND PINS

Baffling to the unenlightened, the symbols on the patches and pins adorning a biker's jacket (and also club membership cards and tattoos) have specific meanings. The diamond-shaped "1%" patch indicates the wearer considers himself worthy of being counted among the outlaw fraternity. Club members may also wear an officer's patch—"president," "vice president," or "sergeant-at-arms"—indicating an official position within their club. Worn by the Hell's Angels, the "AFFA" patch is a symbol of loyal brotherhood, the letters standing for "Angels Forever, Forever Angels." Other clubs wear their own versions—the Outlaws, for instance, wear "OFFO" and the Bandidos "BFFB."

The "81" patch is worn exclusively by the Hell's Angels, the figure representing the eighth and first letters of the alphabet—"H" and "A"." These days "81" is more generally seen gracing the T-shirts and baseball caps of the club's supporters.

The "Filthy Few" patch is controversial. According to the media and many cops, it indicates the wearer has killed for his club, but outlaws claim this is sensationalist nonsense. The patch, they argue, denotes a club member of hefty status—the term dating back to halcyon days when outlaws who were the first to arrive at a party and the last to leave dubbed themselves "the filthy few."

OTHER COMMON PATCHES AND SYMBOLS

The figure "13" on a diamond-shaped patch was usually worn on the breast pocket, on the opposite side to the "1%" patch. It's often thought the "13" patch was a coded way of announcing the wearer as a marijuana user—"M" being the 13th letter of the alphabet. Back in the 1950s and 1960s, this might have been true. More recently, however, the patch simply indicates the wearer subscribes to the outlaw's "outsider" ethos. Traditionally unlucky, the number 13 is embraced as a challenge to conventional taboos.

The number "69" on a circular background appeared on jacket patches during the 1960s. Its sexual connotations were obvious, and the patch was worn for its shock value. Today, however, it's seldom seen.

Iron crosses, swastikas, and Nazi regalia were a common sight on outlaw jackets during the 1960s and 1970s—usually adopted for their shock value. The swastika is now rarely seen, though the iron cross has enjoyed a renaissance and appears on many biker insignia and logos. Used self-consciously, it shows an awareness of the symbol's subcultural heritage.

Rarely appearing these days, "DFFL" patches and pins were a celebration of drugs—"Dope Forever, Forever Loaded."

Still widely seen, especially in the US, the legend "FTW" celebrates the outlaw's rejection of mainstream attitudes and routines—"Fuck the World."

Run patches and badges indicate the wearer's participation in various runs and rallies. Such events are central to biker culture, so a significant number indicates the wearer is a high mileage, well-traveled veteran. "RIP" patches and tattoos are also common, worn in memory of a lost brother.

Iron Cross Rebels

Since the late 1940s motorcycle outlaws had worn iron crosses and other German military regalia, and during the 1960s swastikas and Nazi motifs were common. Brought home as war trophies by GIs, iron crosses were originally worn as a source of victorious pride. The swastika was also used "to show some class." The press was hungry to furnish readers with lurid tales of wicked bikers, and often spotlighted their use of the swastika. The outlaws, for their part, were quick to appreciate the symbol's shock value. As Hell's Angels president, Sonny Barger, explained to a *Los Angeles Times* reporter during the 1960s, "This stuff—iron crosses, the Nazi insignia, the German helmets—that's to shock people. To let them know we're individualists. To let 'em know we're Angels… Hell, we buy this junk in dime stores."

Nazi regalia was often part of the biker's image—but rather than a badge of political belief, it was usually worn for its shock value.

Hog Wild Choppers

Just as the outlaw's appearance announced his status as a breed-apart, so too did the style of his bike. The stripped-down ethos of the 1940s and 1950s survived into the 1960s, but the minimalist "bob-job" mutated into the chopper. The chopper dispensed with all the heavy parts of the factory-produced "garbage wagon." In their place went begged, borrowed, or stolen replacements, as the chopper-builder crafted a more unique machine.

Crafting the Custom Chopper

Razor-sharp cool was the chopper's trademark. Extended front-forks gave more stability on long, Californian highways and—set off by high-rise "ape hanger" handlebars—they also made for a distinctly mean-looking machine. The chopper's footrests and controls were placed right at the front, allowing a relaxed, laid-back riding position, while the rigid, no-suspension frame positioned the rider close to the ground so he almost sat in the bike rather than on top of it. Exhaust pipes soared skyward and smaller "peanut" gas tanks were often cannibalized from Sportsters or British bikes. High-backed "banana" or "king and queen" seats were topped off with backrests (known as "sissy bars") often crowned with attention-grabbing crests—skulls, iron crosses, and daggers. And a dash of glamor came through bright paintjobs and lashings of chrome plating.

In the early days, custom parts couldn't be bought over the counter, so building a chopper required much skill and ingenuity. Gradually, however, small "chop shops" appeared, offering services like tuning, welding, spraypainting, and chroming—even the creation of handcrafted, unique specials.

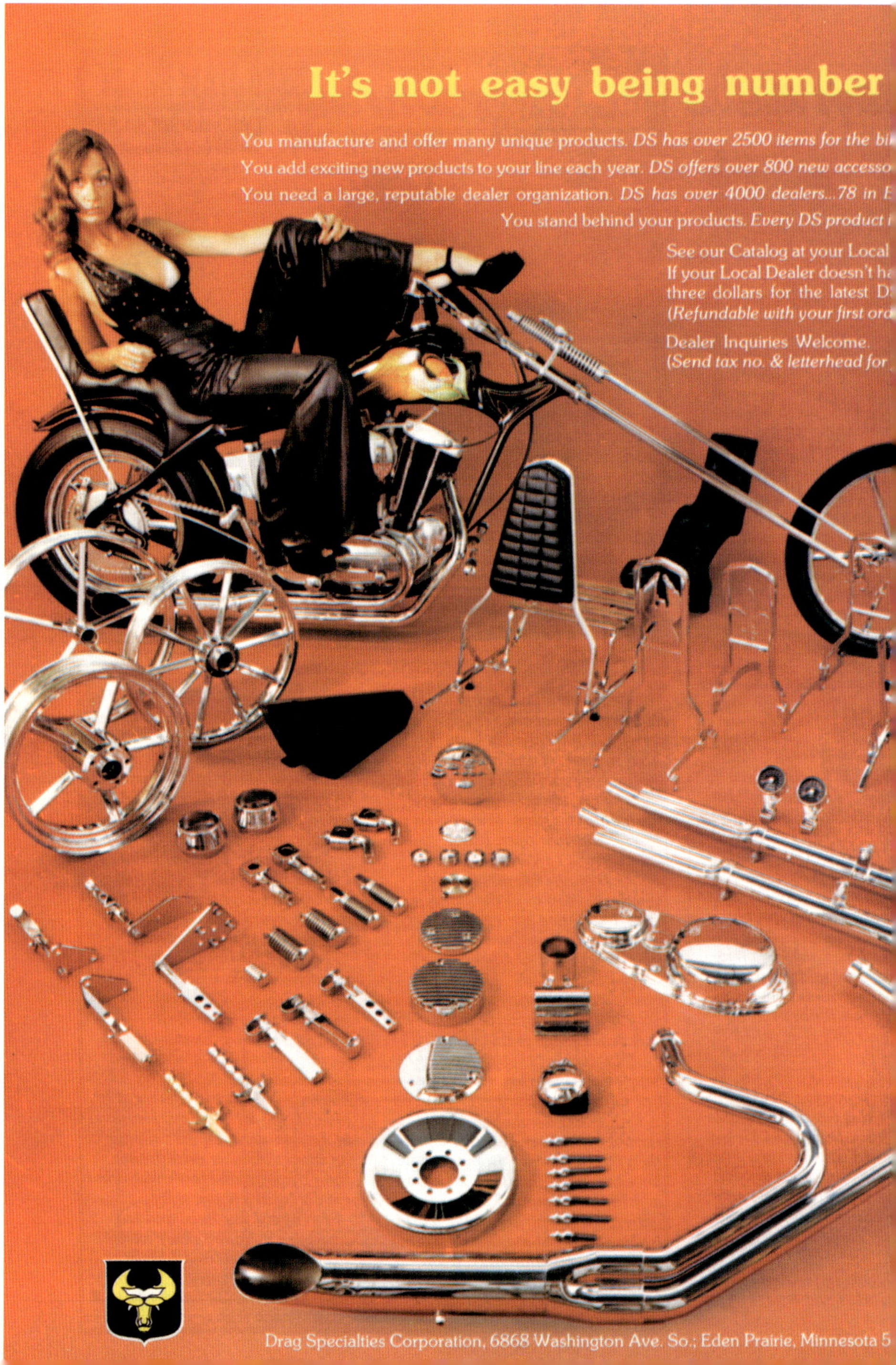

With the growing popularity of choppers, the production of custom bike parts became big business.

The manufacture of custom parts also became big business. In 1967 Tom McMullen launched a motorcycle division of Automotive Electrical Engineering (AEE), a firm that he originally founded to produce custom components for performance cars. Dozens more like-minded entrepreneurs followed his lead. Jammer Cycle Parts, Drag Specialties, Custom Chrome, and Gary Bang all developed into major companies. Components supplied by McMullen and the rest were crucial to the growth of the chopper subculture, while the gospel was spread by a new generation of maverick biker magazines.

Born to Be Styled

As the custom-bike business flourished, and chopper designs became more expressive, custom motorcycle shows were organized (many by outlaw clubs) where bike builders could compete for coveted prizes. By the mid-1960s show-bikes had developed into a recognizable art form, with well-known builders such as Ed Roth, Ron Simms, and Arlen Ness achieving cult status.

Increasingly outlandish, showbikes became virtually unrideable during the late 1960s and early 1970s. Psychedelic and pop art symbolism became a major influence—with bikes developing outrageously extended front-forks, acres of glistening chrome, and gleaming metalflake or candy-colored paintjobs. By the late 1970s, however, the gaudy excess was losing its appeal. Instead, there was a profusion of esthetic styles. The traditional outlaw principle of lean looks and high performance became especially influential, initiating a trend toward simple, clean (but always imaginative) styling, and a return to the classic "bobbed" looks of the 1940s and 1950s.

BIKER LIFESTYLE MAGAZINES

Magazines played a pivotal role in popularizing the low-riding cool of the chopper. During the late 1960s the mainstream motorcycle monthlies shunned the custom scene for fear of alienating their advertisers and traditional readership. Instead, a new breed of more streetwise biker mag emerged to fill the gap in the market.

Despite being a cult hit, Ed "Big Daddy" Roth's *Choppers* magazine (launched in 1967) had folded by 1969. *Street Chopper*, however, proved more enduring. Launched in 1969 by Tom McMullen, at the outset the magazine was virtually an "advertorial" for AEE (McMullen's custom parts firm), but it developed into a forum where hundreds of chopper builders and parts' manufacturers showcased their wares.

Like Roth's *Choppers, Colors* was another short-lived classic. Launched in 1970 by Phil Castle, a New Jersey biker, *Colors* was oriented to the outlaw scene and featured photos of club members, their bikes, and their girlfriends. But distribution was poor as some newsstands refused to carry it. As a consequence *Colors* folded in 1971.

Easyriders was more successful. Like *Street Chopper, Easyriders* started life as a sideline for a custom parts manufacturer—Jammer Cycle Parts. Run by chopper fanatics Joe Teresi, Lou "Spider" Kimzey, and Mil Blair, the firm produced its own, biannual *Jammer's Handbook*. A densely illustrated catalog of Jammer's products, the *Handbook* featured technical articles, cartoons, and bikini-clad models. A hit with customers, the catalog's success spurred the launch of a full-blown magazine "for the swinging biker"—and *Easyriders* was born in 1971. In 1972 Keith Ball joined the magazine's staff and steered its expansion into a $35 million-a-year company encompassing 14 magazines, 10 annual motorcycle and tattoo events, and a comprehensive line of franchise products.

A posse of other US titles followed *Easyriders* trail—*Iron Horse, Outlaw Biker, Biker Lifestyle*, and many more, while similar magazines were launched in countries across the world.

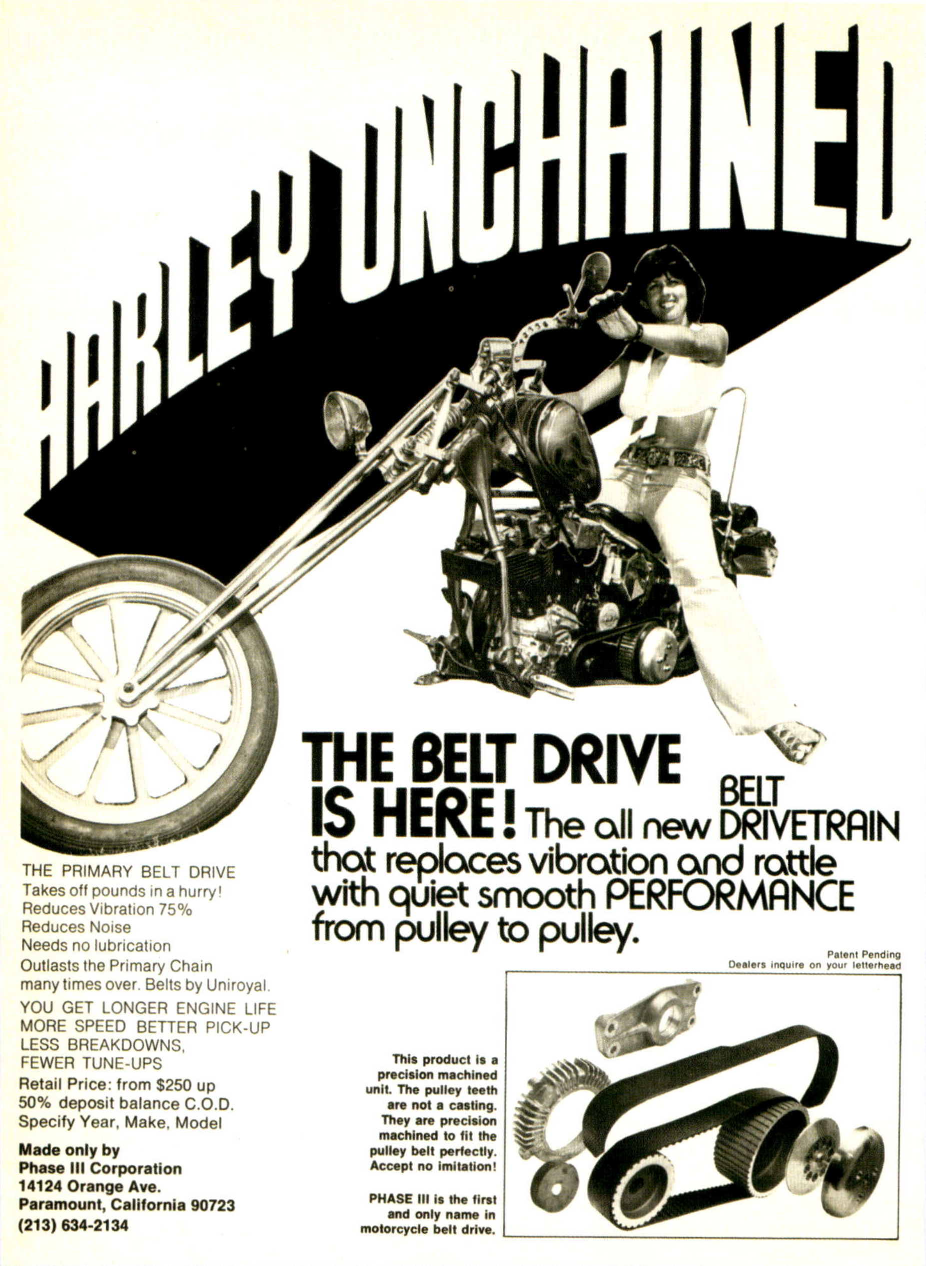

The "Factory Custom"

The original outlaw clubs had tolerated a smattering of British bikes among their ranks, though they had always preferred homegrown Indians and Harleys. But during the 1960s attitudes hardened. Indians retained curiosity value, but it was the Harley-Davidson—invariably a hulking 74—that became the outlaw's bike of choice.

Harley-Davidson's executives, however, were nervous of the outlaw faithful. Throughout the 1940s, 1950s, and 1960s the company held the outlaw image at arm's length and maintained a reputation for conservative respectability. During the 1960s, Harley was preoccupied with the Japanese "invasion," and launched small, fun bikes to compete. In 1966, meanwhile, the Panhead engine (Harley's big-bike powerhouse) was replaced by the "Shovelhead"—so-called because the engine's rocker covers resembled the back of a coal shovel. The Shovelhead was intended to improve acceleration, but it disappointed many chopper riders who felt the engine's weight made their bikes lethargic.

Harley-Davidson's staid approach to design got a major shake-up with the appointment of William G. Davidson as Vice President of Styling in 1969. Willie G. (as he is better known) was the grandson of the company's founder, and he ushered in a new, innovative era that saw the firm increasingly acknowledge the outlaw heritage. A particular milestone was the FX Super Glide, introduced in 1971. The first of Harley's "factory custom" models, the Super Glide was a production bike that incorporated "custom" features such as buckhorn handlebars, rakish forks, and low-slung frame. A sea-change in Harley's corporate strategy, the Super Glide heralded a long line of outlaw-inspired "factory customs" that included the Bad Boy, the Fat Boy, and the Heritage Springer.

Probably the most recognizable motorcycles in the world were the choppers ridden by Peter Fonda and Dennis Hopper in *Easy Rider.* The originals were stolen just before the film's release, but faithful replicas have been produced by several custom bike firms, including Panzer.

"EASY RIDER" AND CLASSIC CHOPPERS

Following the release of *Wild Angels* in 1966, choppers became the staple ingredients in biker movies. But it was *Easy Rider* that elevated the chopper to a popular icon. Audiences were wowed not only by the footloose lifestyle portrayed by Peter Fonda and Dennis Hopper, but also by the daring, defiant look of their motorcycles—Fonda's ultra-raked, "stars-and-stripes" cruiser and Hopper's orange-and-flame lowrider.

Before shooting began in 1968, Fonda purchased four Harley Panheads at a Los Angeles Police Department auction. To prepare the bikes for their movie roles, they were chopped, welded, and spraypainted—and during shooting they were kept on the road by Fonda, stuntman Tex Hall, and actor Dan Haggerty. As a precaution, four bikes were built so the crew had substitutes if there was a break-down. In the event, all four held up to the rigors of moviemaking—though one was sacrificed in the movie's final crash scene.

Fonda was an experienced motorcyclist, so his bike was more radically customized than Hopper's. True to the chopper ethos, Fonda's hog dispensed with all the factory model frills—gone were the front fender, the suspension, and the sprung seat. Added were stretched front-forks, "ape hanger" handlebars, and swept-up, "fishtail" exhaust pipes. Eye-catching chrome was everywhere, while the stars-and-stripes paintjob on the gas-tank underlined the movie's search for an authentic America.

Less at home in the saddle, Hopper's ride was more moderately chopped. Rather than soaring "ape-hangers," the bike's risers brought the minimalist drag bars close to the rider. A small gas tank replaced the larger stock model, while a frame-hugging seat gave the bike sleek, clean lines—set off by the licking flames of its custom paintjob.

The bikes were priceless pieces of motorcycle history but, before the movie's release, the three undamaged machines were stolen. Never recovered, it's likely the bikes were broken down and sold off as parts by thieves unaware of their potential value.

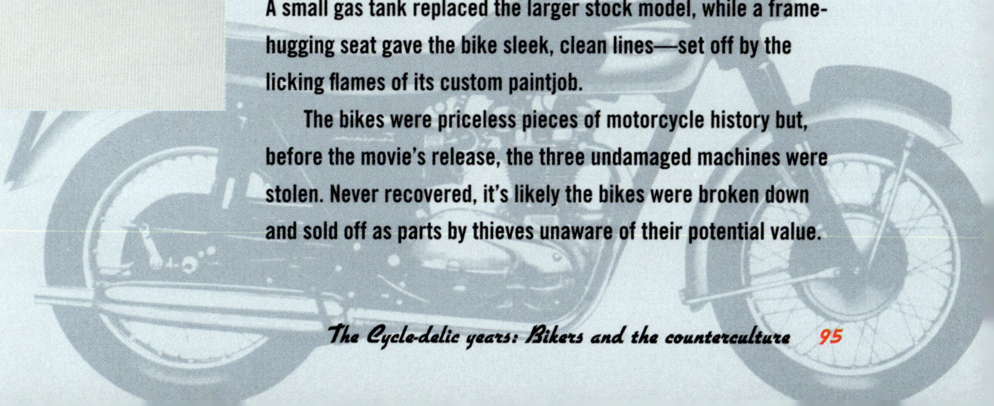

Let It Bleed

The East Bay scene was a volatile cocktail of factions, movements, and subcultures, where tensions quickly bubbled to the surface. Bikers and hippies both defied mainstream norms, but in many ways the blue-collar outlaws were a world away from the predominantly middle-class bohemians—and the escalation of the Vietnam War increased the tension.

Allen Ginsberg speaking from a step ladder in London's Hyde Park on the legalizing of pot.

"I Ain't No Commie"

While on the one hand, bikers were staunchly patriotic and prowar, many hippies and students generally opposed the war in Vietnam. By the fall of 1965 the University of California campus at Berkeley was a hotbed of political radicalism and passionate opposition to the Vietnam War. On October 16th Allen Ginsberg, hippy insurgent Jerry Rubin, and beat poet Gary Snyder organized a Vietnam Day Committee (VDC) march to protest against the war. From the Berkeley campus, the 1,500 demonstrators planned to march to the Army Terminal near downtown Oakland—but the protestors were halted just outside the city by a phalanx of riot police. Then, as the cops looked on, a squad of around a dozen Oakland Hell's Angels suddenly ripped into the crowd. Led by Sonny Barger, the Angels tore up placards and laid into the demonstrators with fists, boots, and chains—one massive biker reportedly hollering "By God, I'm an American an' I ain't no Commie."

The police finally intervened, and in the ensuing mêlée six Angels were arrested. But, for once, the establishment was sympathetic. Young Republican Clubs commended the Hell's Angels for their patriotism and Fred Ulner, director of the pressure group Republicans for Conservative Action, announced he was collecting donations to pay off the Angels' fines. In the event, however, it wasn't necessary. All the outlaws escaped prosecution, save one Angel who walked out of court with a paltry $56 fine.

"A Group of Loyal Americans"

Stunned by the Hell's Angels' attack, the grandees of psychedelia scurried to patch things up with their erstwhile allies. With another march planned for November, Ginsberg quickly organized a summit meeting between the radicals and the Angels. Ginsberg serenaded the bikers with Buddhist chants and told a bemused Sonny Barger that he loved him. Pacified by copious amounts of grass, the bikers grudgingly agreed to cease hostilities. With their finely tuned instinct for publicity, the Oakland Angels called a press conference and announced they wouldn't intervene in the next demonstration. But they still had nothing but bile for the protestors, and explained to reporters they were staying away from the march "because our patriotic concern for what these people are doing to our great nation may provoke us to violent acts." Furthermore, as a neat parting shot, Sonny Barger fired off a telegram to the then President, Lyndon B. Johnson:

Dear Mr. President,
On behalf of myself and my associates,
I volunteer a group of loyal Americans
for behind the lines duty in Viet Nam.
We feel that a crack group of trained
gorillas would demoralize the Viet
Cong and advance the cause of freedom.
We are available for training and duty
immediately.
Sincerely,
Ralph Barger Jnr.
Oakland, California
President of Hells Angels.

Altamont

Their Satanic Majesties

President Johnson never took up the Oakland Angels' offer. The friction between the outlaws and the hippy radicals, however, was unmistakable. An uneasy truce spluttered until 1969, but the counterculture's veneration of the maverick biker was finally exploded by the debacle of Altamont.

The Rolling Stones' 1969 US Tour had been a phenomenal success. But a few rankles had been prompted by high ticket prices so, as a "thank you" to America, the Stones planned a free concert in San Francisco. Originally the idea was to stage the event in Golden Gate Park, but the deal fell through. Under pressure to find a quick alternative, the concert was finally set at Altamont Speedway, southeast of Frisco. There was little planning or foresight, but the Rolling Stones were confident the show would be a sensational success.

More than 300,000 people showed up for the free event, planned as a daylong festival featuring Santana; Crosby, Stills, Nash, and Young; The Flying Burrito Brothers; Jefferson Airplane; Ike and Tina Turner; and The Grateful Dead. In the past The Dead had hired the Hell's Angels as security, and the Stones had used them at their free concert in Hyde Park without a problem, so the club seemed a natural choice as security at Altamont. But maintaining order around the hurriedly assembled stage—hardly 4-feet (1.5-meters) high—was an impossible task. The show was tense from the start, and sporadic violence increasingly punctuated the day as fractious Angels struck out at an edgy and impatient crowd.

At Altamont in 1969 the Hells Angels' weapons of choice were cut-down pool cues.

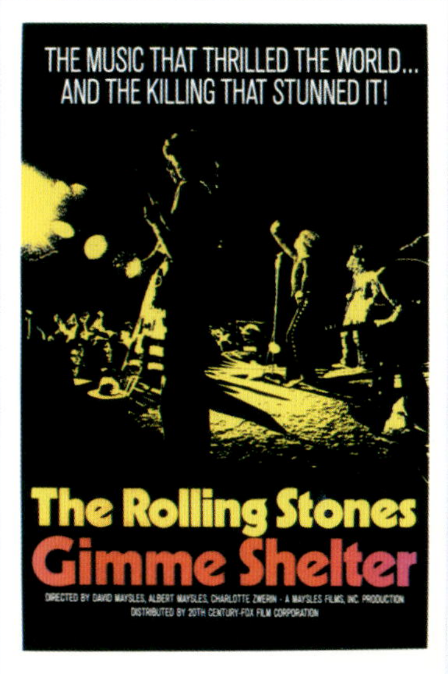

Released in 1970, *Gimme Shelter* chronicled the carnage at Altamont. Footage from the movie was also used in the successful defense of a Hell's Angel charged with the murder of Meredith Hunter.

The Angels' choice of weapon for the day was cut-down pool cues, the bikers beating senseless anyone who got in the way. Even the performers were fair game, and in a scuffle near the stage a burly Angel knocked Jefferson Airplane's Marty Balin unconscious. As dusk fell, it seemed like the Angels were running amok.

As on the rest of the tour, the Stones built up tension by waiting as long as possible before making their entrance. But, by the time the group were on stage, all hell was breaking loose. As the band kicked into "Sympathy for the Devil," an ugly confrontation began between the Angels and Meredith Hunter, an 18-year-old black kid. When Hunter pulled out a gun, the bikers steamed in, beating and stabbing him to death. The whole sorry saga was caught on camera by David and Albert Maysles, hired to chronicle the Stones' tour in a movie ultimately released as *Gimme Shelter* in 1970.

A San Francisco Hell's Angel named Allen Passaro was later charged with Hunter's murder. Pleading self-defense, Passaro was acquitted because the Maysles' movie footage clearly showed that Hunter had pulled a gun. Nevertheless, the Altamont carnage had finally exploded the counterculture's naïve worship of the outlaw biker.

Al Adamson's *Satan's Sadists* led the way in a new wave of ultra-sleazy biker exploitation movies.

Satan's Sadists

For exploitation moviemakers, Altamont was a gift. Their murderous reputation confirmed, outlaw bikers featured in a new spurt of violent biker pics such as *Angels Die Hard* (1970), *Hell's Bloody Devils* (1970), and *Under Hot Leather* (aka *The Jesus Trip*, 1971). There were even attempts to capitalize on the Altamont violence, with the hippy commune replacing the small rural town as the target of biker carnage in exploitation fodder such as *Angel Unchained* (1970) and the brutal *Angels, Hard as They Come* (1971). But all were eclipsed by *Satan's Sadists* (1970). Promising "Human Garbage—In the Sickest Love Parties!", the movie was the baby of schlock-movie master, Al Adamson. In *Satan's Sadists* Adamson's bikers—the Satans—are rotten, sons-of-bitches who leave a trail of rape and murder through the Mojave Desert. Adamson's no-budget shocker also took advantage of the media uproar surrounding the murders committed by the Charles Manson family in summer 1969. Manson, himself, had shared the counterculture's awe of motorcycle outlaws. But, the bikers thought Manson was a weirdo. And it was a member of the Straight Satans biker gang who finally tipped off detectives that Manson had been bragging about his clan's murderous trail.

Soul Bikers

Outlaw bikers of the 1960s wore swastikas for their shock value, but their Nazi insignia won them a reputation as racists. And, during the 1970s, 1980s, and 1990s, white supremacist prison gangs such as the Aryan Brotherhood were known to number many bikers among their membership. Often, however, this membership was just an expedient way of surviving in the violent, dog-eat-dog world of America's prisons. Certainly, individual outlaws may have held racist views, but racism was never a defining feature of the biker creed.

During the 1960s Hell's Angels chapters had an ambivalent relationship with local African-Americans. Oakland, for example, was home turf to both the Angels and the Black Panthers (the paramilitary group of black radicals), but the two camps kept their distance. California's black biker clubs, on the other hand, were on pretty good terms with the Hell's Angels.

Outlaw Outsiders

In a sense, black bikers were outlaws from the start. During the 1920s and 1930s African-Americans were excluded from AMA affiliated clubs, so black clubs developed outside the motorcycle mainstream. Black motorcycle clubs of the 1940s and 1950s such as the Berkeley Tigers and the Star Riders (with chapters in Oakland and Los Angeles) tended to attract older guys, who played it straight and rode full-dresser motorcycles. The 1960s, however, saw the emergence of black bikers who were much closer to full-on outlaws.

Oakland's East Bay Dragons started out as a 1950s car club—but in 1959 they cashed-in their cars for bikes and sewed their old car club colors onto the back of greasy Levi cut-offs. Modeling themselves on the Hell's Angels, the Dragons always looked mighty mean on their chopped-down Harleys. Other black clubs also sprung up around Oakland—the Headhunters, the Wicked Wheels, the Zodiacs, and the Choppers. The Funky Wheels were down in San Jose, while over in Los Angeles the biggest, baddest black clubs were the Defiant Ones and the Chosen Few.

Maori clubs such as the Mongrel Mob are a major force in New Zealand's outlaw biker scene.

As well as African-Americans, other ethnic groups have also been attracted to the outlaw biker lifestyle. In New Zealand, for example, Maori bikers became a major presence in the outlaw scene from the late 1960s onward, with the rise of Maori clubs such as the Stormtroopers, Black Power, and the Mongrel Mob—named (reputedly) after a 1950s Court Judge said that a group of Maori men before him were "nothing but a pack of mongrels." The "Mighty" Mongrel Mob took a belligerent bulldog as their colors and have since grown to a network of more than 30 chapters throughout New Zealand.

Many Native Americans have also been drawn to the world of the biker—the subculture offering a sense of power and identity to a range of ethnic groups alienated from the social mainstream.

Brotherly Love

Press reports of the 1960s often speculated that outlaw bikers might be gay. For the most part, this was pretty wide of the mark. The idea probably originated from the media's misunderstanding of the outlaws' habit of deep kissing one another. Rather than having a sexual motive, the ritual was partly a demonstration of fraternal loyalty, and partly a mischievous attempt to "freak out" the citizens.

But there's no mistaking the way the myths and rituals of biker subculture have had a strong homoerotic appeal. Since the 1950s, the macho world of chrome and leather has found resonance in gay subcultures, and biker imagery has figured prominently in the catalog of gay style.

Scorpio Rising

Incendiary moviemaker Kenneth Anger was quickly drawn to the homoerotic charge of the biker look. Anger—an occultist, perverse visionary, and premier figure of American avant-garde cinema—was based in Paris for most of the 1950s, where he completed *Hollywood Babylon*, a biography of Tinsel Town's sordid underbelly. Returning to New York in 1963, Anger continued his exploration of society's seamy underside. Running across a motorcycle gang, Anger was inspired to make the movie that made him America's foremost independent moviemaker—*Scorpio Rising*.

Premiering in New York in October 1963, *Scorpio Rising* was later described by Anger as "a death mirror held up to American culture." Stylized and iconoclastic, *Scorpio Rising* made explicit the edgy homoeroticism of biker culture, the picture intercutting visceral images of motorcycle subculture with a medley of media allusions. A sardonic commentary on American mythologies of glamor, power, and machismo, the movie featured muscled bikers striking fetishistic poses in boots, buckles, and studded leather.

When *Scorpio Rising* played in San Francisco in 1964, the Hell's Angels went to check it out. Although they liked the movie, many were angry about the way the cinema had used press clippings about their club to promote it. As one annoyed outlaw explained:

> ***We all enjoyed it. But then we came outside and saw all those clippings about us, pasted up like advertisements. Man, it was a bummer, it wasn't right. ...Shit, did you see the way those punks were dressed? And those silly goddam junk-wagon bikes? Man, don't tell me that has any connection with us. You know it doesn't.***

***Scorpio Rising*'s mix of polished chrome and black leather was an anthem to homoeroticism.**

05

SISTERS IN LEATHER

Women and Biker Culture

Women in the Saddle

Motorcycle culture drips with machismo. The testosterone-charged world of the biker is often characterized by Stone Age attitudes and macho postures, with women invariably relegated to the pillion seat. In advertising spreads and biker mag centerfolds, women have been objectified as foxy, leather-clad sirens, or as voluptuous sex kittens draped around the gleaming chrome of a massive, phallic machine.

Posing on the back of a Henderson bike, Mack Sennett's "Bathing Beauties" pushed risqué as far as it would go during the 1920s.

The chauvinism of the biker creed has often obscured the significant role women have played in the world of motorcycling. Over the years many women have been enthralled by the mobility and glamorous adventure offered by the motorcycle. During the 1880s bicycles provided the same sense of emancipation, and many women took to two wheels in defiance of Victorian conventions of femininity. Motorcycles magnified the liberating appeal. With their widening availability during the early twentieth century, motorcycles provided the modern, independent woman with an unrivaled opportunity to shatter an array of gender barriers.

Even during the motorcycle's earliest days, women had enthusiastically jumped into the saddle. Pioneering women riders included 18-year-old Clara Wagner who, in 1910, braved driving rain and mud-drenched roads to clinch a 365-mile (584-km) endurance race from Chicago to Indianapolis—though the patriarchs at the Federation of American Motorcyclists (the predecessor to today's AMA) refused to recognize her victory. Avis and Effie Hotchkiss were also trailblazers. The mother-and-daughter team completed a 5,000-mile (8,000-km) transcontinental roundtrip in 1915, running their Harley-Davidson sidecar rig from New York to San Francisco and back again. In 1916 two sisters, Adeline and Augusta Van Buren (descendants of former President Martin Van Buren) also completed a transcontinental trek—this time astride two Indian Powerplus bikes. Outstanding riders, the sisters negotiated rocky trails and treacherous deserts.

The "Queen of Miami"

Bessie Stringfield was another early biking heroine. Known as the "Motorcycle Queen of Miami," Stringfield was a petite African-American woman from Florida. Fiercely independent, she began riding in the late 1920s, when it was unheard of for a black woman to be seen on a motorcycle. Stringfield couldn't even get a bike license until a police officer interceded on her behalf. Soon she became a skilled stunt rider and made eight solo cross-country rides, later serving as a dispatch rider with the US Army. During the 1950s Stringfield founded the Iron Horses Motorcycle Club and in 1990, at the age of 79, she was honored by the AMA at the opening of their Heritage Museum in Ohio. Stringfield died in 1992 at the age of 81, and in 2000 the AMA created an award bearing her name—to be presented to women riders who have shown Stringfield's brand of pioneering spirit.

Known as "The Motorcycle Queen of Miami," Bessie Stringfield began riding in Florida during the 1920s—a time when it was unheard of for an African-American woman to be seen on a motorcycle.

Ladies of the Highway

Dot Robinson carried the torch for women motorcyclists during the 1930s. Co-owning a Harley-Davdison franchise with her husband, Robinson wore "ladylike" pink riding outfits and even fixed a lipstick holder to her pink-painted bike. But she could also be as tough as leather. A regular competitor in grueling endurance races, in 1940 she won the classic Jack Pine Endurance Run—a punishing, two-day race through Michigan.

Robinson was also a leading light of the Motor Maids, America's first national club exclusively for women riders. The idea for the Motor Maids was originally conceived by Linda Dugeau, a motorcycle enthusiast from Rhode Island, who envisioned the group as a motorcycling equivalent to the 99 Club —an elite fellowship of women pilots, established in 1930 by the famous aviator, Amelia Earhart. Dugeau wrote to motorcycle dealers and riders up and down the country, finally tracking down 51 women who became the charter members of the Motor Maids. Founded in 1940, the group were the ultrarespectable face of women's motorcycling.

With Dot Robinson as their president (a position she held for 25 years), the mission of the Motor Maids was to unite women riders in the promotion of motorcycling. Since the 1940s the Maids have held meetings and get-togethers throughout the United States. With around 500 members in chapters all over America, the Motor Maids maintain a regular presence at many motorcycle events, as well as hosting their own annual convention—where a motorcade of members parade in the group's neat uniform of gray slacks, blue blouses, and white gloves, boots, and ties.

Biker Chic

Sexuality and motorcycles are inextricable. The erotic charge of chrome and black leather is electrifying, while the sexual connotations of straddling the massive cylinder blocks of a Harley are unmistakable. Unsurprising therefore is the eroticized pairing of beautiful women and powerful machines—a theme that has been played out to the full in popular culture.

Even during the 1920s and 1930s, the sexually provocative combination of sexy women and enormous bikes was picked up by *Film Fun* and other saucy magazines of the day. The theme also surfaced in Hollywood. Posing on the back of a hefty Henderson bike, Mack Sennett's "Bathing Beauties" (the slinky artistes used as decorative extras in the movie producer's slapstick comedies) pushed risqué as far as it would go.

Cycles and the "60s Sizzle"

The "Swinging Sixties" saw the motorcycle become a de rigeur fashion accessory for the smoldering starlet. The raunchy machine was an essential complement to big hair, a bigger cleavage, and a sultry pout. The Swedish-born star, Ann-Margret, became famous for indulging her wild side on a Harley. Hailed as the "female Elvis," in 1964 she starred alongside the King himself in *Viva Las Vegas*, and went on to feature in dozens of 1960s movies, TV specials, and Vegas stage shows. Ann-Margret was a passionate biker and frequently appeared astride a motorcycle, both on screen and in real life. In 1970 she even starred alongside gridiron legend Joe Namath and "Big" Bill Smith in the biker movie *C.C. and Company*. A spill from her bike in August 2000 cost her some broken ribs, but the actress was unfazed and remains a devotee of two wheels.

Sixties sex kitten, Brigitte Bardot, also paired up with a motorcycle. A media sensation through her daringly independent lifestyle, in 1967 the "princess of pout" provided the vocals to Serge Gainsbourg's steamy love anthem, "Harley-Davidson."

Marianne Faithfull got in on the act too. The face of swinging London, Faithfull delved into the eroticism of leather and chrome in the movie *Girl on a Motorcycle* (1968). Then Mick Jagger's girlfriend, Faithfull played a restless French newlywed who hits the road on her bike in an odyssey of self-discovery. Zipped into her fur-lined, leather jumpsuit (and nothing else), the nymphet is shown speeding through the European countryside, her thoughts lost in a reverie of sexual fantasy. Based on André Pieyre de Mandiargue's novel, *La Motocyclette*, the movie actually amounted to little more than a series of arty vignettes.

Slinky sixties starlet Ann-Margret was a diehard motorcycle fan.

and even Chanel conceived styles that took their cues from the Harley-riding hardcore. Biker chic roared off the catwalks of Paris and New York into the pages of classy magazines such as *Vogue* and *Harper's Bazaar*.

The motorcycle leathers in the trendy fashion pages conveyed strength and power. But this was often undercut by frilly skirts and revealing tops, flawless makeup, and models who struck poses of passive availability. Nevertheless, biker chic was the order of the day, and high-profile personalities lined up to be photographed in motorcycle mode. Wynonna Judd, Queen Latifah, and k.d. Lang all paraded their "biker" credentials, appearing in glossy photospreads decked out in leathers and straddling their own machines. The huge cachet of biker style even persuaded Harley-Davidson to cash in on the trend, and in 1989 the firm launched a new division—Harley-Davidson MotorClothes. The range was slick and stylish, though many of the "bikers" who bought Harley's outfits didn't actually own a motorcycle. Biker chic hadn't made it fashionable to be an outlaw—merely to act the part.

In 1967 Brigitte Bardot straddled a massive chopper for a movie clip promoting Serge Gainsbourg's torrid lovesong, "Harley-Davidson."

From the Back Streets to the Catwalk

The doyennes of high fashion also became infatuated with biker style. Inspired by the motorcyclist's air of rebellion and cocksure cool, trendy designers of the 1980s and 1990s created what became known as "biker chic"—an upmarket collection of women's fashions that blended battered denim, chrome buckles, and body-hugging black leather. Famous designers like Donna Karan, Anne Klein,

Ready to Ride

The torrid blend of high-powered motorcycles and voluptuous women also became a well-established theme in contemporary erotica—from softcore pinups and video collections such as *Penthouse* magazine's "Ready To Ride," to the skin-and-chrome centerfolds that are a staple ingredient of biker lifestyle mags.

She-Devils on Wheels

The classic biker pics featured many of the same conventions and stereotypes as softcore porn. Often, women were presented as vulnerable and submissive sex objects who fall prey to marauding sexual predators. Skin-flick king Russ Meyer, for example, plundered the imagery of the malevolent biker in one of his earliest movies, *Motorpsycho!*

Other biker pics were less ironic. Indeed, political correctness never figured highly on the biker movie's agenda—attested to by the casual regularity with which women were raped and molested in movies such as *Devil's Angels* (1967), *Born Losers* (1967), *Angels Die Hard* (1970), and *The Tormenters* (1971). Living up to its name, *Sleazy Rider* (1973) was especially lurid. A near-hardcore porn outing, it features the (decidedly chubby) Conchos Motorcycle Club—a gang of greasy, chopper-riding fatboys, with a taste for open-air orgies and gang-raping the local sheriff's daughter.

The Biker-Chick Pics

There was, however, also a biker pic subgenre that turned two-wheeled machismo on its head. In fact the late 1960s and early 1970s saw hog-hauling, ass-kicking biker chicks become a major draw on the drive-in circuit. In *The Hellcats* (1967), for example, fearsome femmes run riot across the Mexican border. The *Mini Skirt Mob* (1968), meanwhile, features a pack of toothsome biker chicks known as The Mini-Skirts, who terrorize young newlyweds.

In *Sisters in Leather* (1969) things get more salacious (not to say a little weird), with a gang of badass, lesbian bikers blackmailing a business executive, and seducing his straight-laced

The Hellcats was one of many biker pics that featured hog-hauling, ass-kicking biker chicks.

wife into their wicked ways. Another posse of feisty biker femmes also take to the road in Angels' *Wild Women* (1972). Tackling rapists and devil worshipers along their not-so-easy ride, the gang terrorize any man who dares cross their path. The biker vixen at the center of *Bury Me an Angel* (1971) also shows men who's boss. Played by the statuesque, 6-foot (1.8-m) tall Dixie Peabody (her real name), the shotgun-toting Amazon hits the highway with two compadres, looking for revenge against the low-life SOB who shot her brother.

Man-Eaters on Motorbikes

With their no-budget production values, threadbare plots, and wooden acting, the biker chick-pics all followed in the glorious traditions of exploitation moviemaking. Probably the greatest example of the oeuvre is *She-Devils on Wheels* (1968). Independently produced by Herschell Gordon Lewis (who had earlier crafted the schlock-horror epics *Blood Feast* (1963) and *Two Thousand Maniacs* (1964), *She-Devils on Wheels* was an over-the-top classic. The film centers on The Man-Eaters—a sadistic, chain-wielding, all-girl biker gang who select their sexual playthings from a "stud line," later dragging the discarded beaus behind their speeding hogs. *She-Devils on Wheels* is an outrageously camp slice of 1960s trash. The Man-Eaters sport colors of a pink, bow-tied pussycat, while the gang's brazen attitude is summed up in the movie's catchy title song "Get Off the Road":

We are the hellcats nobody likes
Man-Eaters on motorbikes
We own this road so you'd better get lost
When you hear the roar of a cut-out exhaust

Running With the Pack

With some justification, the outlaw biker subculture has a reputation as a man's world. But this hasn't always been the case. Back in 1947, for example, the *San Francisco Chronicle* reported that at the Hollister fracas, of the raucous crowd "perhaps one out of ten was a girl, wearing slacks and a tight sweater." And, during the late 1940s and early 1950s, the chapters of some outlaw clubs—for example, the San Francisco and San Bernardino Hell's Angels—had a smattering of women members. Many others also rode along with the clubs, some astride their own hulking machines.

Mamas, Sheep, and Old Ladies

By the 1960s, however, women were barred from official membership of outlaw motorcycle clubs and were consigned to the pillion seat, both literally and metaphorically. Stories of outlaws' abuse of women abound. And it was a key theme in Attorney General Lynch's 1965 report on the "menace" of outlaw motorcycle clubs. Subsequently repeated as gospel truth in newspapers, magazines, and movies, the Lynch Report stated with clear-eyed conviction that the women who consorted with biker gangs fell into three categories—"mamas," "sheep," and "old ladies."

"Old ladies" were the wives or long-term girlfriends of club members. Out of bounds to other bikers, they were regarded as the property of their guys and were generally expected to be seen but not heard. The notion of the "sheep" was more outrageous. With an air of apparent authority, the Lynch Report explained how the Hell's Angels' club rules specified that every prospective member was required to take with him to his "initiation ceremony" a compliant woman or "sheep"—to be offered as a ritualistic sexual sacrifice to his brother bikers. The story was reiterated in lurid magazine stories, but was entirely spurious. The Angels, however, were quite happy to have people believe it, taking perverse delight in thinking that straight society believed them to be anti-social barbarians whose lifestyle was way beyond the pale of conventional morality.

The concept of "mamas" was grounded more firmly in reality. "Mama" was a term used to describe the girls who hung around with an outlaw club, but weren't attached to any particular member. In his account of the Hell's Angels' history, Hunter S. Thompson explained that "mama" was all that remained of the expression "Let's go make somebody a mama," and was used to denote female camp-followers who were sexually available to any outlaw or favored guest. According to Thompson, each outlaw club had a handful of mamas, with the Oakland Angels having as many as five or six at any one time. In the outlaw subculture, mamas were regarded as fair game for anyone. But they weren't always pathetic dupes. Some could be as wild and aggressive as the men of their chosen club.

A big feature in the outlaw scene of the 1960s, mamas have since largely disappeared. Partly this is a consequence of club members getting older. Inevitably, many outlaws got married and had kids, and their old ladies can't have been best pleased with the idea of free-and-easy mamas hanging around their old man's club.

"Property Patches" are still seen around—though how seriously they are taken by their wearers is debatable.

Property Patches

In the past, some outlaw clubs went so far as to give their women "Property Patches." Declaring that the wearer was the "Property of" a designated club (for instance, "Property of Outlaws"), the Property Patch was the size of a full set of colors usually worn on the back of a mama's or old lady's cut-off. Satan's Slaves took things a step further, marching their womenfolk down to the tattoo parlor to have "Property of Satan's Slaves" imprinted permanently on their rears.

Today, however, the practice is almost universally considered passé. Indeed, even when the Property Patch was in style, how seriously it was actually taken is debatable. Together with the wearing of swastikas, Nazi helmets, colored wings, and other offensive motifs, the Property Patch was often just an attempt to "show some class" by blowing the minds of the regular citizens.

The Motorcycle Sisterhood

Since the 1960s women have had an unmistakably subordinate role in the subculture of outlaw bikers. Outside the male-dominated outlaw clubs, however, women were keen to mark out their own exclusive biker territory.

Women on Wheels

The number of motorcycle organizations catering for women has multiplied since World War Two. Alongside the ultrarespectable Motor Maids there exists an array of other training-oriented groups and social clubs geared to women riders. With a worldwide membership of over 2,000, the Women's International Motorcycle Association (WIMA) is probably the largest women's motorcycle organization in the world. Its first International President, Louise Scherbyn, founded the group in America in 1950. Her husband (himself a keen motorcyclist) encouraged her to learn to ride during the 1930s, and Scherbyn was soon a passionate biker and the proud owner of a 1932 Indian Scout. During World War Two Scherbyn began corresponding with women riders throughout the world and came up with the idea for an international organization for women motorcyclists. There now exists a global WIMA network, with branches throughout North America, Europe, Australia, and Japan. Drawing its membership from diverse social backgrounds, the group exists as a forum for women bikers to share their experiences, problems, and ideas.

Established in 1979, Women in the Wind (WITW) exists to "unite women motorcyclists with friends of common interest, and promote a positive image to the public of women on motorcycles." The group is smaller than the WIMA, but now has over 1,000 members in chapters located in the US, Canada, Great Britain, and Australia. Women On Wheels (WOW) shares a similar vision. Founded in California in 1982, its mission is to represent the interests of women motorcyclists, providing a sense of communal solidarity for its mainly American membership.

Outlaw Biker Babes

All-woman motorcycle clubs proliferated during the 1980s and 1990s. Founded in 1983, the family-oriented Leather and Lace MC now has chapters in Vermont, Pennsylvania, Massachusetts, New York, and Connecticut, and works to promote safe riding and to provide support for women and children's charities. Other clubs are more dedicated to the social life of rallies, runs, and hard riding. The East Coast Biker Chicks MC, for example, is based in Boston. Helkats MC are based in the Midwest, while Washington State has the B.A.D. Girls MC, and in Washington DC there are the Speed Divas. In Santa Cruz there's Sweet Evil WMC, while the Priestess Motorcycle Covern draws members from North Carolina, Indiana, and Pennsylvania. Lesbian riding clubs have also formed to advance their members' lifestyles. Best known among them are San Francisco's Dykes on Bikes, while in New York there are the Sirens and in Boston the Moving Violations MC. Confident and strident, the new wave of women's motorcycle clubs have forged a biker sisterhood whose camaraderie and riding skills easily match that of the macho outlaws.

Some women's clubs even adopt the subcultural paraphernalia associated with the outlaw hardcore—leather, denim, large capacity bikes, and full sets of colors. Based in San Francisco, the Devil Dolls have made a particular media splash. Formed in 1999,

No longer confined to the pillion seat, modern women bikers are every bit as cool and confident as their male counterparts.

the self-proclaimed "outlaw biker babes" are lithe, leather-clad girls who share a passion for powerful Harley hogs. Image-conscious and media-literate, their carefully crafted public image of hard-riding, independent, and sexy femmes won them an hour-long PBS TV special and acres of coverage in the biker press.

The Future is Female

Women riders have also caught the attention of Harley-Davidson, and now represent a crucial sector of the company's market. A decade ago women accounted for only two percent of Harley's sales, but now the figure stands at nine percent and is set for further growth.

As a way of establishing closer links with its female customers, Harley-Davidson launched the Ladies of Harley (LOH) in 1986. LOH was a spin-off of the Harley Owners Group (HOG), established by the company to generate a sense of community and brand loyalty among its new customers (who gain membership automatically). The Ladies of Harley was launched as a way to encourage HOG's female members to become more involved in chapter activities, and LOH are always a big presence at HOG rallies and runs. Harley have even developed bikes with the women rider in mind. Introduced in 1988, the lightweight 883cc Sportster Hugger's seat height was a modest 26 inches (66 centimeters), allowing even the most petite rider to master the bike. The strategy paid off and, by the mid-1990s, nearly 24 percent of Harley Sportsters were being bought by women.

Times have undoubtedly changed since the unrepentant sexism of the 1960s motorcycle scene. Today women bikers are hardly a novelty. In the US, women now represent over eight percent of motorcycle riders—and the proportion looks set to grow in the coming years.

UPJ 749
95CYY

06

OUTLAWS WORLDWIDE

A Global Brotherhood

The Coffee-Bar Cowboys

The mean, moody, and magnificent biker made his presence felt throughout the world during the 1950s and 1960s. In Europe, the economic growth of the postwar era made youth a new consumer force, and young people used their economic muscle to assert their idealism and independence. For many kids, the allure of American popular culture—ducktail haircuts, blue suede shoes, rock 'n' roll—appeared like a bold and exciting Promised Land compared to the drab, old-fashioned world of their parents.

During the postwar years the image of the biker began to spread a special charisma over much of Europe. Glimpsed in American magazines, advertisements, and movies—most obviously 1954's *The Wild One*—the tough, leather-jacketed road warrior promised an unsurpassed sense of freedom and empowerment. European kids lapped up the motorcycle rebel's heroic swagger, and they were soon mimicking his attitude, his style, and his disrespect for authority.

Transplanted to Europe, the biker's surly machismo won few admirers in officialdom. The establishment hated the audacity of the new generation and, across the continent, there were sharp confrontations between kids and the forces of orthodoxy—and it was the irrepressible biker who stood out as the intrepid rebel par excellence. In Austria and West Germany the halbstarken (or "half-strong," as motorcycle tearaways in leather jackets were known) were blamed for a new spirit of lawless defiance among the young. In Sweden it was the skinnknutte or "leather-jackets," and in France the blousons noirs or "black jackets." But it was in Britain that the biggest brigade of black leather speedsters took shape.

The "Ton-Up Boys": Bikers in Britain

Rather than "bikers," Britain's motorcycle desperados of the 1950s and 1960s were known as "Ton-Up Boys," "Coffee-Bar Cowboys," or, later, "Rockers." In Britain the immediate postwar years were an austere time, with gas rationing only ending in 1950. But during the 1950s the good times began to roll, and kids enjoyed a newfound affluence. Increasing numbers could afford their own motorcycle—a scruffy, second-hand Triumph, Norton, or BSA Twin, or even a spanking new machine bought on the "never-never" of retailers' new credit schemes.

For the British Ton-Up Boys, like other speed junkies around the world, the motorcycle was more than a means of transportation. It was a statement of outlook, a resolute assertion of lifestyle that offered escape from the dead-end monotony of work and home. The image of the American wild ones added extra attitude to the adventure. Screenings of *The Wild One* were banned in Britain until 1968 (the censorship adding to the biker's illicit aura), but the iconography crossed the Atlantic in movie stills and magazine features, and was eagerly adopted by Britain's young bike crowd. The Trailmaster and Barbour waxed-cotton jackets traditionally favored

The British "Wild Ones"—the Ton-Up Boys take to the road in 1964.

by British riders remained much in evidence, but increasingly it was denim jeans and Brando's black leather look that took hold.

For a while the BSA Gold Star—affectionately known as "the Goldie"—was the British biker's ride of choice. An early version of the Gold Star had appeared in 1938, but sales were slow and the war put an end to production. Returning in peacetime, the classic versions of the Gold Star appeared in 1954 as the 350cc CB32 model and the 500cc CB34. But, as in the States, British riders were seldom satisfied with stock bikes. Increasingly, modified machines replaced the Goldie as the premier British racer. Like the American "bobbers" of the 1940s and 1950s, British bikers chopped away their machines' superfluous fittings and bolted on racing fuel tanks, swept back pipes, and low, "clip-on" handlebars—all for the aim of "doing the ton"—tipping the speedo above 100 mph (160km/h).

Speed Demons.

The Ton-Up Boys' nightly "burnups" didn't go unnoticed. During the early 1960s the police began to take a tougher line against the reckless "menace." As statistics for motorcycle accidents soared, the media also waded in, and in 1961 a "shock issue" of the *Daily Mirror* announced that "on British roads today, the only vehicle carrying more dead people than the motorcycle is a HEARSE."

The Café-Racer Years

For Britain's Ton-Up Boys, life revolved around the coffee-bar, café or "caff." Pubs and bars of the time rarely welcomed leather-jacketed bikers with open arms—they tended to upset the older clientele and attracted unwanted attention from the local constabulary. The "caff," on the other hand, was where it all happened. Warm and inviting (and, best of all, cheap), the "caff" became the regular haunt of many British riders.

It was a place where they could meet up with mates and spend smokey evenings "chatting up the birds" and spinning a few disks on the jukebox. Some cafés became legendary biker hangouts—the Nightingale in South London, the Salt Box near Biggin Hill, the Chelsea Bridge Tea Stall, the Busy Bee on the A41, Johnsons on the A20, and (most famously) the Ace Café on London's North Circular Road.

A café racer revs up his Royal Enfield bike outside The Sportsman's Rest café in Kent in 1960.

Record Racing

The jukebox opened the way for the most infamous of the café racer sports—record racing. The stunt involved slapping a record on the jukebox—maybe a single by Eddie Cochran, Buddy Holly, or Gene Vincent—and leaping onto your bike in an attempt to complete a prearranged course before the end of the song. A popular route from the Ace covered a distance of 3 miles (5 km). Massed races from the Ace to the Busy Bee, 12 miles (20 km) away on the Watford bypass, were another café-racer favorite.

Outside the Ace Café, Bill Shergold (the "Ton-Up Vicar") and young bikers watch the evening's action.

THE ACE CAFÉ

The legendary Ace Café was a "greasy spoon" transport café (or diner) on London's North Circular Road. Built in 1938, its formica-topped tables, and chairs screwed to the floor, were intended for truck drivers. But during the late 1950s and early 1960s the Ace became a bikers' Mecca.

Open 24 hours a day, the Ace became the Ton-Up Boys' favorite pitstop. It was the place to get together, have a cup of tea, arrange runs, or simply tinker with your bike. In its heyday, a busy evening could see up to a 1,000 bikes come in and out of the Ace. After midnight, when the North Circular traffic was at low ebb, breakneck "burnups" could be held with little fear of interference. But a growing number of accidents brought increased police patrols—although, to catch the faster bikes, the local constabulary had to bring in a powerful V8 Daimler sports-car.

Sadly, the Ace closed its doors in 1969 and the building became a tire-fitting depot. But regular reunions were held during the 1990s, the 1997 event attracting over 25,000 riders. And in 2001 the Ace Café was resurrected. Driven by a passion for fast bikes and rock 'n' roll, Mark Wilsmore secured ownership of the original Ace site. Lovingly, he completely restored the buildings and has put the celebrated Ace Café back in business as a regular meeting place for bikers old and new.

The Leather Boys was filmed around the famous Ace Café and featured many local bikers as extras.

Ton-Up Movies

The Ton-Up Boys were big news in the late 1950s and early 1960s. The action and atmosphere of the café racer scene even appeared in a handful of British movies. *The Damned*, produced in 1961 (but not released until 1963) was probably the first movie to feature British bikers. An unusual science-fiction tale, *The Damned* features Oliver Reed in his first major role as "King," the cold-blooded leader of a pack of motorcycle thugs. *Some People* (released in 1962) also touched on biker themes—but tells the unlikely story of a group of teenage loafers persuaded to form a pop group as an alternative to hanging around with wayward bikers. More convincing was *The Leather Boys*. Adapted from Gillian Freeman's novel (first published under the pseudonym Eliot George in 1961), *The Leather Boys* is a claustrophobic melodrama, distinguished by its harsh realism and sexual frankness. But it was also notable for its locations. Shot against the backdrop of London's budding motorcycle subculture, *The Leather Boys* was filmed around the famous Ace Café—with local riders and their bikes drafted in as extras.

Yikes! *The Damned* featured Oliver Reed (in the pale jacket) as the vicious leader of a British motorcycle gang.

THE TRITON

The Triton was the classic bike of the café-racer era. In the profusion of homemade, custom-styled machines, a common arrangement was a powerful Triumph engine (usually a 500cc or 600cc Twin) slotted into a Norton Featherbed frame. The combination offered the best of both worlds—a reliable engine that delivered a fair turn of speed, and a lightweight frame that allowed excellent handling and road-holding. So commonplace was the Triumph/Norton fusion that it became known as a bike in its own right—the "Triton." No two Tritons were the same. Every conceivable permutation of parts—racing fuel tanks, clip-on handlebars, custom exhausts—were tacked onto the basic Triumph/Norton foundation. Many, however, shared the classic Triton paintwork of black cycle parts and silver tank.

By the mid-1960s, the Triton and its ilk were being eclipsed by the rise of powerful sports bikes from overseas. The Italians had always produced machines in the café-racer tradition, but during the mid-1960s firms such as Laverda, Moto Guzzi, and Ducati brought their flair to the big-bike market, producing impressive, factory-bred racers. And during the 1970s Japanese firms like Honda, Yamaha, and Suzuki refined the formula, producing some awesome machines tailor-made for the speed-hungry.

"Rev in Leather"—Reverend Bill Shergold poses outside the Ace Café.

The mighty Triton was the classic bike of Britain's café-racer era.

THE 59 CLUB

During the early 1960s the famous 59 Club grew into one of the largest motorcycle organizations in the world. Established by Reverend Bill Shergold, it began as a general youth club in East London's Hackney Wick. Father Bill was already a keen motorcyclist—progressing from a Bantam to a succession of Triumph Speed Twins for traveling around his parish—and he soon hit on the idea of starting a club specifically for young bikers. Encouraged by the Triumph Owners Club, Shergold drummed up support among the regulars at the Ace Café and other ton-up haunts. And in October 1962 the 59 Club was relaunched as a motorcycle association with a membership of several hundred.

With his dog collar just visible beneath his scarf and bike leathers, Reverend Shergold was a dependable presence among his two-wheeled flock. The 59 Club was first and foremost a motorcycle club, but Shergold did his best to counsel the young members. The motorcycle Vicar regularly visited injured riders in hospital and spoke up in court for those that had gone astray. Shergold's endeavors attracted considerable media interest, and even won backing from the Bishop of London—who cheerfully came down to bless the 59 Club's breakdown truck.

Beachside Battles: Mods and Rockers

The British media rarely regarded young motorcyclists with favor. But press coverage hit a nadir in response to the pitched battles between "mods" and "rockers" that flared up during 1964. The episode began at the seaside town of Clacton-on-Sea, southern England, where a cold, wet Easter holiday saw scuffles between local youths and visiting Londoners.

The disturbance was fairly minor—a few windows were broken, some beach huts vandalized, and a few dozen kids on scooters and motorbikes roared up and down the promenade. But, in the absence of other newsworthy material, reporters from national newspapers seized upon the fairly innocuous events and created headlines suggesting there had been wholesale mayhem. In a panic uncannily similar to the American hysteria that surrounded 1947's "invasion" of Hollister, the British media spoke of a shocking "day of terror" that had seen a sleepy town ransacked by young gangs "hell-bent on destruction."

The Who's *Quadrophenia* recreated the thrills and spills of the mods 'n' rockers era.

The Seaside "Invasions"

The overwrought press reports conjured with images of an all-out battle between two distinctive youth clans—the "mods" and "rockers." The rockers were the leather-clad, rock 'n' rolling Ton-Up Boys. The mods were a new subculture of clean-cut sharp dressers who favored Italian scooters and cool American soul. In the wake of the Clacton fracas, 1964 saw further sensationalized media coverage of Bank Holiday (civic holiday) "battles" between mods and rockers in Hastings, Margate, Brighton, and other seaside resorts across southeastern England, and the stories continued sporadically throughout the mid-1960s.

In reality, however, the incidents were often wildly exaggerated by the press. The media coverage, in fact, actually helped escalate events. The mods and rockers had initially been fairly ill-defined youth styles, but were given greater form and substance in the sensational news stories. And the two subcultures steadily polarized as youngsters throughout Britain began to identify themselves as members of either camp—the mods or the rockers.

The Boys Are Back in Town

By the late 1960s, however, the mods were mutating into something new. A harder-edged version of mod style gradually coalesced into the distinctive "uniform" of the skinhead—steel toe-capped boots, rolled-up jeans, suspenders, and closecropped hair.

The rockers, meanwhile, were influenced by the American outlaws. Increasingly known as "greasers," they grew their hair longer and traded in their café racers for choppers. Forming themselves into loose-knit clubs, a number adopted the name "Hell's Angels" (the only outlaws they'd heard of) and emulated their US cousins by sewing crude replicas of the infamous death's head insignia onto the backs of grubby denim cut-offs.

THE MODS

Inspired by chic continental design esthetics, the mods were Britain's slick, stylish youth subculture of the early 1960s. The smooth lines of Italian fashion were first sported in Britain during the late 1950s by the "modernists"—hip, West London cliques. As the "modern" look caught on, the kids of blue-collar workers adopted light, finely tailored (yet tastefully understated) mohair suits with short jackets and tapered trousers, the ensemble kept pristine beneath a GI's green, fishtail parka. And, rather then a dirty motorcycle, the mods' preferred mode of transportation were the zippy Italian scooters manufactured by Vespa and Lambretta—often modified with swanky paintjobs, fur trim, extra headlights and mirrors, and tons of chrome.

Popular mod haunts were nightclubs such as the Scene and the Flamingo where they danced all night (fueled by "Purple Heart" amphetamine pills) to West Indian ska and bluebeat music and, especially, American soul and rhythm and blues (the latter emulated by groups such as The Who and The Small Faces). Stylish and clean-cut, the mods were often fêted by the media as pacesetters of the "Swinging Sixties." At the same time, however, they could also be reviled as the bête noire of affluent society, bent on causing mayhem.

QUADROPHENIA

The 1960s' glory days of the mods and rockers were recreated in the movie version of The Who's rock opera, *Quadrophenia*. Shot on location in Brighton, the film focused on the life of a young mod, Jimmy (Phil Daniels), and his search for meaning amid the seaside battles of 1964. Released in 1979, the success of *Quadrophenia* contributed to a revival of mod style in Britain during the late 1970s and early 1980s.

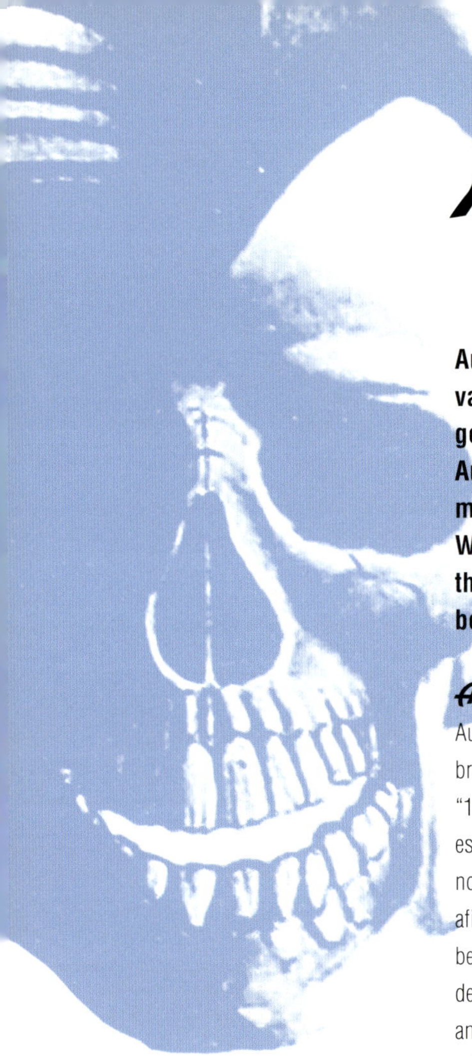

Bikers Down Under

Australia also embraced the attitude, style, and values of the biker—or the "bikie," as he is generally known down under. Like Britain's rockers, Australia's bikies of the 1950s and early 1960s took many of their cues from Brando's leather-jacketed Wild One. By the end of the 1960s, however, it was the outlaw clubs of America's West Coast who had become the leading inspiration.

Aussie Outlaws

Australia has proved fertile soil for the development of outlaw biker brotherhoods. One of the first Aussie bike clubs to embrace the "1%er" identity was the Gladiators, established as early as 1963. As the notoriety of American outlaws spread far afield, US clubs such as the Hell's Angels became a key influence on the development of Australian bikie culture, and homegrown clubs increasingly mimicked the image and organizational structure of their American brethren.

The late 1960s saw countless single-chapter motorcycle clubs spring up throughout New South Wales (especially around Sydney) and Western Australia (around Perth). Some clubs also boasted several chapters dotted around the vast continent. The Finks, for example, were established in Sydney during 1969, and soon had additional chapters in southeast Queensland, Adelaide, and Wodonga. The Rebels were also founded in Sydney back in 1969. During the 1990s they developed into Australia's largest outlaw motorcycle club, with 63 chapters and over 1,000 members.

The Gypsy Jokers was yet another club established in Sydney in 1969. Originally a homegrown Aussie club, within a few years of formation the Australian Jokers had affiliated to the American club of the same name and had established further chapters in Perth, Adelaide, Newcastle, and Mount Gambier. The Gypsy Jokers also developed close links with other clubs—notably the Fourth Reich, which was founded during the early 1970s. Later, hassle from the police prompted several Fourth Reich members to relocate to Queensland, where they established a new club called the Black Uhlans. Other outlaw clubs included the Coffin Cheaters, Odin's Warriors, and the Commancheros.

American clubs also staked a claim down under. The worldwide growth of the Hell's Angels during the early 1970s saw the first Australian chapters of the club inaugurated in Melbourne and Sydney in 1973. But during the 1980s the Angels began facing tough competition from the Bandidos, a rival US club who had also set their sites on international expansion. Dissident members of the Commancheros in Sydney established the Bandidos' first Australian chapter in 1984, and by 2005 as many as 12 Bandido chapters existed.

Stone **is the classic Australian "bikie" movie.**

"Stone"—The Film and Its Following

When Australian bikies talk about movies that reflect their lifestyle, they may give a nod of recognition to *The Wild One* and *Easy Rider*, but it's more likely they'll cite *Stone* as the movie that, more than any other, encapsulates the style and attitude of bikies. A huge success on its original release in 1974, *Stone* was absorbed into Australian biker folklore and to this day retains a faithful following.

The movie was the brainchild of law-school dropout Sandy Harbutt. Riding a powerful Triumph, Harbutt hung around the fringes of bikie subculture during the early 1970s, and became fascinated with the bikies' rebellious spirit, their camaraderie, and fierce sense of loyalty. At the time, the Australian government was trying to kickstart a national movie industry, and Harbutt managed to secure financial backing from the Australian Film Development Corporation for *Stone*—the first Australian biker movie.

Written, produced, and directed by Harbutt, *Stone* is a thriller with a simple, if unlikely, plot. After an outlaw motorcycle club—The Gravediggers—witness a political assassination, its members begin to be murdered one by one. To investigate the killings a young drug squad cop (Stone, played by Ken Shorter) infiltrates the outlaw subculture, but gradually "goes native" as he immerses himself in The Gravediggers' way of life. The implausibility of the plot, however, is unimportant, because what made *Stone* a cult classic was the way Harbutt handled his material.

The producer/director tried to tell the story from "the inside," striving for a faithful representation of the Aussie outlaws' lifestyle and lore. Harbutt's attention to detail and passion for the bikie culture are evident throughout his movie. Along with members of the Hell's Angels and Lone Wolf M.C., recruited as extras,

Stone **derives its sense of gritty realism from the many authentic "bikies" drafted in as extras.**

the atmospheric locations, moody camerawork, and killer soundtrack also helped intensify the movie's gritty appeal.

On its release *Stone* was panned by critics but was a hit with audiences. The movie struck a particular chord with the bikie subculture it depicted, and over the years has won devoted fans among Australian outlaws. *Stone*'s loyal audience kept the movie alive, and in December 1998 34,000 bikies descended on Sydney to mark the movie's 25th anniversary.

Kamikaze Riders

Japan's first motorcycle gangs roared into life during the early 1950s. As in Europe, Japan's postwar affluence delivered a new spending power into the hands of the nation's youth. As a consequence, Japanese cities saw the emergence of new youth styles and subcultures, dubbed zoku or "tribes" by the media. Many of the young zoku borrowed styles from their counterparts in the West, especially the US. Leather jackets and jeans, for example, were typically favored by Japanese bikers who became known as the kaminari-zoku—or "thunder tribe"—on account of their bikes' raucous exhausts.

A Japanese biker shows off his embroidered tokkofuku—the calling card of the bosozoku.

Birth of the Bosozoku

Although the Japanese press complained about noisy motorcycles and the occasional traffic accident, throughout the 1950s and 1960s the kaminari-zoku weren't considered much of a problem. But the image of motorcycle gangs changed during the late 1960s and early 1970s following several widely reported episodes of violence. One of the most serious was 1972's Toyama Jiken (Toyoma Incident), when a motorcycle gang led 3,000 rioters in a fierce battle with the police. In 1974 and 1975 a series of clashes between rival motorcycle gangs in Kanagawa (a state close to Tokyo) attracted more media attention. But, above all, it was the disturbances that took place in Kobe in 1976 that guaranteed the media's vilification of Japanese bikers. The confrontation began when police tried to stamp out an illicit race meeting held by an assembly of local motorcyclists. Events took an ugly turn, and a crowd of several thousand went on the rampage, wrecking patrol cars and stoning a police station. The Kobe incident became notorious and gave currency to a new term—bosozoku. The name was used to describe the new tribe (or zoku) of sharply dressed young bikers whose lives were dedicated to the excitement of boso—high-speed, high-risk riding through Japan's hectic city streets.

Emperors of Speed

The mid-1970s through to the early 1980s were the heyday of the intrepid bosozoku or "speed tribe." Mostly in their teens or early twenties, bosozoku gangs such as Black Emperor, Hell Tribe, and the Fierce Tigers transformed the city streets into their playground. On summer weekend nights Japan's congested cities exploded with the din of souped-up motorcycles as tens, often hundreds, of bosozoku assembled to show off their machines and stage impromptu races through the crowded metropolis.

Like the American outlaws, the bosozoku had a taste for the outrageous. Their motorcycles were modified virtually beyond recognition, with bland factory models transformed into breathtaking kaizosha (modified vehicles) that reflected the owner's personal style. While US outlaws favored hulking 1200cc Harleys, the bosozoku generally opted for more-readily available Suzukis and Yamahas; and the strict Japanese traffic laws meant they were generally limited to bikes of around 250–400cc. But the diehard bosozoku always stamped his personality on his machine. Frames were resprayed in bright, primary colors, gas tanks were painted with enigmatic Chinese characters (chosen for their visual appeal), while an array of custom mirrors, fenders, sissy bars, and

headlights added style and attitude. For some, the original purchase price of the motorcycle was nothing compared with the total costs of modification.

In appearance, the well-groomed and carefully dressed bosozoku were a far cry from America's greasy chopper crew. In place of leathers and denims, the bosozoku of the 1980s favored tokkofuku. Literally, tokkofuku means the uniform of the kamikaze, but in contrast to the dull, utilitarian overalls of the original suicide pilots, the bosozoku's elegant outfits were finely tailored in shades of yellow, pink, and white as well as black and gray, and were painstakingly embroidered with group names, Chinese symbols, and phrases suggesting strident nationalism—for example, "Patriot" or "Protection of the Nation and Respect for His Majesty." Similar slogans also decorated headbands embroidered with the Imperial rising sun crest, which was also resplendent on the flags held aloft by young hatamochi (flag holders) who rode pillion.

The nationalist symbolism, however, didn't point to a close relationship between the bosozoku and Japan's political right wing. On the contrary, the bosozoku had little respect for nationalist movements, and used the symbols for their shock value. Just like the US outlaws, the bosozoku loved to "show some class," with displays calculated to infuriate the ippanjin (ordinary citizens).

In their heyday, hundreds of bosozoku would parade through Japan's hectic city streets.

Sogo Ishii's manga-influenced *Burst City* (1982) led a posse of hi-octane, Japanese biker movies.

The Japanese biker movies of the 1970s and 1980s, such as *Crazy Thunder Road* and the *Detonation!* trilogy, were raw-edged and action-packed.

The Bosozoku Roadshow

Throughout the early 1980s bosozoku were regarded as one of Japan's major social problems. In reality, however, there was always a marked element of playful theatricality to the bosozoku. They reveled in medatsu, or "being seen." Flaunting their uniforms and slogans, and executing wild riding stunts on Japan's busiest streets, the bosozoku relished acting out the part of heroic, freedom-loving speedsters—giving a thrilling, high-octane performance that blasted away the boredom and conformity of mainstream convention.

The media was also swept up in the spectacle. The press ostensibly deplored the young riders' outrageous behavior, but was nevertheless captivated by the bosozoku's exciting image, and eagerly chronicled their audacious exploits. A sizable commercial industry also developed around the bosozoku. Publishers such as Daisan Shokan produced a torrent of books and magazines brimming with pictures and profiles of bosozoku groups throughout Japan. Rock groups such as the Cools and Yokohama Ginbae (Yokohama Silver Flies) also adopted bosozoku fashions and incorporated bosozoku themes in their songs (complete with sound effects of thunderous bike exhausts).

Japanese bikers also appeared on the silver screen. The "Alley Cat Rock" series of the early 1970s was a succession of exploitation action pics based around the Alleycats—a bad-ass gang of shapely biker chicks. Teruo Ishii's *Bakuhatsu!* (Detonation!) movie trilogy of the mid-1970s, meanwhile, introduced a mob of supercool motorcycle delinquents, the Black Cats. And in the early 1980s Sogo Ishii was acclaimed for his raw-edged, manga-influenced biker pics *Crazy Thunder Road* (1980) and *Burst City* (1982). But the most enigmatic of the bosozoku-era movies was Mitsuo Yanagimachi's 1976 documentary, *God Speed You, Black Emperor*. Following the exploits of the Black Emperor bosozoku clan, the movie chronicles virtually every aspect of its daredevil lifestyle—from the club meetings and races to the showdowns with rival gangs.

The International Outlaws

As biker culture and lifestyle spread worldwide so, too, did the reach of the outlaw motorcycle clubs. During the 1960s and 1970s the global circulation of biker movies, magazine features, and news stories ensured that the image and ethos of America's two-wheeled wild bunch were exported far and wide. As a consequence, indigenous outlaw clubs sprang up throughout the world, imitating the structure and style of the US originals.

Even behind the iron curtain the shaggy-maned, tattoo-adorned motorcycle outlaw made his mark. Formed during the early 1980s, Moscow's Nochnyye Volki, or Night Wolves, became one of Russia's leading outlaw clubs. During the Soviet era, the club was an enemy of the state, and members worked as bodyguards in the rock music underground. But, with the fall of communism, the Night Wolves had an image makeover and turned from antiestablishment subversives into civic-minded entrepreneurs. Staunch supporters of Russian President Vladimir Putin, the Wolves now claim good relations with city governments and even helped build a school outside Moscow. The Night Wolves have also helped launch a Russian counterpart to the Harley-Davidson. Made at the Irbit Motorcycle Factory in the Ural Mountains, the Russian chopper was christened (naturally enough) the "Wolf."

A Worldwide Biker Brotherhood

American outlaw clubs have also hoisted their colors in new territories, granting charters to new chapters in virgin territories across Europe, Canada, and Australia. The Hell's Angels were quick off the mark. During the 1960s the club, under Sonny Barger's leadership, had grown from an ad hoc collection of misfits into a finely tuned national institution with its own organizational structures and hierarchy. With its base of operations firmly established in the States, the Angels began a phase of international expansion. Reputedly, an Auckland Hell's Angels chapter was established in New Zealand as early as 1961, but the club's key period of overseas growth really began during the late 1960s and 1970s when charters were granted in Germany, Switzerland, and England. Further growth followed, with charters granted throughout

Originally founded in 1959, the Pagans developed into one of the East Coast's largest outlaw biker clubs.

mainland Europe, Scandinavia, Australia, and the Americas. And, with the fall of the iron curtain, the Angels' grip stretched eastward, with prospect chapters established in Croatia, the Czech Republic, and (in 2004) Moscow.

Today, the Hell's Angels is a sophisticated multinational conglomerate. Sometimes referred to as the "Big Red Machine" or the "Hell's Angels Motorcycle Corporation" on account of its scale and clout, in the US the club boasts more than 200 chapters in 13 states (mainly on the West Coast and in the northeast). Internationally, a network of more than 200 Angels chapters now snakes across 25 different countries. Outlaw motorcycle clubs are notoriously secretive, but it is believed that the Hell's Angels' global membership currently stands at around 2,000–3,000. The elite status of the club and its close-knit structure means that full-patch membership is highly selective, but each chapter (normally numbering between 10–20 full members) is invariably surrounded by a legion of prospects, associates, and assorted hangers-on.

The "Big Four"

The media uproar that surrounded the Hell's Angels during the 1960s ensured it became the most widely known outlaw motorcycle club in the world—to the extent that the term "Hell's Angel" has been used ubiquitously to describe virtually any leather-jacketed hoodlum on a bike. But the media circus surrounding the Hell's Angels overshadowed the growth of other outlaw clubs, every bit as big and powerful as the Californian-based 1%ers. By the 1990s the outlaw motorcycle subculture was dominated by a quartet of large, international clubs—a big four that comprised the Hell's Angels, the Outlaws, the Bandidos, and the Pagans.

The International Outlaws

The Outlaws

The Outlaws Motorcycle Club, also known as the American Outlaw Association, traces its history back to 1935 when the McCook Outlaws Motorcycle Club was established at Matilda's Bar in McCook, Illinois, just outside Chicago. The original club stayed together throughout World War Two, though its organization became more loose and informal. Order was reestablished during the 1950s as the club grew in size and relocated from McCook to Chicago—its name changing to the Chicago Outlaws in the process. A schism in 1958 saw some members splinter into a racing-oriented club, while others followed truck driver Johnny Davis, the new President of a reconstituted Outlaws.

The Outlaws' colors also underwent change. During the early 1950s the original winged motorcycle insignia was replaced by a small skull, embroidered on the back of black shirts or painted on leather jackets. In 1954 crossed pistons were added beneath the skull, emulating the logo used by Brando's biker gang in *The Wild One*. In 1959 the motif was made larger and more detailed. In 1965, a New Year's Eve party celebrated the formal institution of the American Outlaws Association, and the skull and crossed pistons (christened "Charlie") was taken as the club's grinning calling card—with gothic lettering added to pick out the "Outlaws" moniker.

With the election of a new president in 1984, the Outlaws' center of operations shifted from Chicago to Detroit. A period of expansion followed, and the club now boasts as many as 63 chapters spanning 16 American states (mostly in the East and Midwest) and eight countries (including five chapters in Poland and a prospect chapter in Russia).

The Bandidos

The Bandidos are another global motorcycle club. Originally established in Texas in 1966 by Marine Corps veteran Donald Eugene Chambers, the Bandidos drew many of their early members from disillusioned Vietnam vets. The club even took the Marine Corps' red and gold colors as its own, while the Bandidos' logo—a cartoonish, potbellied Mexican wearing a sombrero and brandishing a pistol and sword in each fist—was originally drawn by a custom car and bike artist by the name of "Lando," who worked around east Houston.

Don Chambers served as the Bandidos' president for six years, but in 1972 he was imprisoned after being accused of involvement in the murder of two bikers in El Paso. Released during the early 1980s, Chambers became an elder statesmen within the Bandidos until his death in July 1999—hundreds of bikers paying their respects at a huge funeral in Pasadena.

The Bandidos now represent a force to be reckoned with. The club remains concentrated in Texas and the Southwest, but now boasts an American membership of over 1,000 in 81 chapters stretching across 15 states—supplemented by 80 support clubs. In Europe, meanwhile, the Bandidos have a membership of over 1,500 across 12 countries—a global reach that closely rivals that of the Hell's Angels.

The Pagans

Originally founded in Maryland during 1959, the Pagans have grown to 44 chapters, mostly spread between New York and Florida. Unlike their rivals, the Pagans don't have a geographically fixed mother chapter, but have a "mother club" made up of former chapter presidents who sport a black number "13" on their colors to indicate their special status. The Pagans are the only outlaw club among the "Big Four" without international chapters, though they have ties to Canadian clubs and (allegedly) have close links with the Mafia.

Biker Wars

By the late 1960s the major outlaw clubs had laid claim to their respective home turf—the Outlaws in the Midwest; the Bandidos in the Southwest; the Pagans on the East Coast; and the biggest club, the Hell's Angels, on the West Coast and parts of the South. A smaller club, the Mongols (originally established as a Chicano prison gang in East LA during the early 1970s) had also staked a claim to parts of southern California. For a while the boundaries seemed well delineated. But during the 1970s things began to change. As the Big Four expanded, rivalries intensified and subsequent decades saw bloody biker wars flare with murderous regularity—not only in America, but across every continent where the outlaw creed had taken root.

The Hell's Angels were still a pretty rough and ready outfit in 1965—but by the 1980s they had developed into a well-oiled organizational machine.

07

GUNNING FOR TROUBLE

Bikers at War

The Good, The Bad, and The Ugly

The outlaw biker of today looks a very different animal from the riders of the 1950s and 1960s. There's still no mistaking the "screw-the-world" attitude, but the oily grunge has been replaced by a much sharper image of classy bikes, custom leathers, and magnificent tattoos. The outlaw clubs have also polished up their reputations with charity fundraising, food giveaways, and support for Vietnam veterans.

The outlaw image is more popular than ever before. The Hollywood gloss has won the "bad boy" style a vast following of weekend rebels who relish taking a walk on the wild side, donning leathers and bandannas, and hitting the road on shiny new Harley-Davidsons. The outlaw image has even been adopted by serving members of the US military and police, who have formed their own backpatch motorcycle clubs, taking names such as The Enforcers, Blue Steel, and The Wild Pigs. Today the biker's negative connotations are almost forgotten and he is celebrated as an all-American hero, a rough-and-ready free spirit who embodies time-honored values of liberty, loyalty, and proud independence.

Of course, most motorcycle clubs exist simply to provide leisure and camaraderie for their members. According to the law enforcement agencies, however, the outlaw hardcore have become sophisticated criminal syndicates. For the cops, outlaw bikers aren't just lovable rogues who enjoy blasting their bikes down the freeway. Beneath the romantic myth, the police maintain the outlaw clubs have developed into a global criminal underground. Based in heavily fortified clubhouses, more like military bunkers, and running extensive networks of associates and subordinate clubs, the outlaws (cops say) are into every criminal racket—from drug dealing and extortion, to prostitution, and money laundering. And, as each club seeks to extend its empire, vicious and deadly gang wars have brought a spiraling body count.

Sonny Barger has retired to Arizona, but he remains a Hell's Angels' elder statesman.

Patch-Overs and Turf Wars

From the early days of the 1950s there were rivalries and rumbles between groups of motorcycle firebrands. But during the 1960s, as the outlaw clubs became bigger and more tightly organized, violent feuds and complex cycles of retribution became an established feature of the outlaw universe.

Searching for greater status and power, the bigger clubs have expanded, swallowing up—or "patching-over"—smaller rivals and absorbing their membership. Sometimes, the process is amicable. Smaller clubs often jump at the chance to drop their own patches and join an outfit with the size and prestige of the Hell's Angels, Outlaws, Pagans, or Bandidos. After a probationary period to prove their worth, the newcomers become a new chapter of the bigger club, adopting its name and colors. Occasionally, however, the takeover is more violent, with the big boys forcibly taking control of the small fry.

The classic biker image of long hair, studs, and greasy leathers steadily disappeared during the 1980s and 1990s as outlaw motorcycle clubs became better organized and more business-minded.

Biker warfare became major news in 1971, as war flared between the Hell's Angels and the Breed—East Coast outlaws with their sights set on expansion. In the Midwest tensions built up as the Breed squared up to the Angels. Battle finally erupted on May 6th at Cleveland's Annual Motorcycle Custom and Trade Show, as around 25 Angels charged into a much larger force of the Breed. Although they were heavily outnumbered, the Angels' attack was ferocious and, of the five bikers that died in the brawl, four were members of the Breed.

The RICO Circus

The turf wars weren't just about winning status and keeping face. During the 1970s and 1980s the control of lucrative drugs operations and other rackets was also at stake. The authorities faced a tough job bringing the outlaws to heel. The clubs' tight organization, discipline, and strict code of silence made them a tough nut to crack. So in the late 1970s the cops resorted to using the Racketeer Influenced and Corrupt Organizations (or RICO) Act.

Passed in 1970, the RICO legislation was originally intended to fight the Mafia, but the Hell's Angels were one of the first RICO targets. A series of prosecutions during the late 1970s and early 1980s sought to convict Sonny Barger and other Angels for racketeering. The trials dragged on for months and cost millions of dollars. Most cases, however, eventually foundered and the small number of convictions were soon overturned after the courts ruled that, while individual Hell's Angels may have committed crimes, the club itself was not a criminal organization. It was a big victory for the bikers, who all along had argued that they were victims of a police witchhunt.

Outlaw Blood Feuds

The authorities' campaign against outlaw clubs continued through the 1980s. Under Ronald Reagan's administration, the 1986 President's Commission on Organized Crime stated that "outlaw motorcycle gangs engage in almost every conceivable crime" and estimated the outlaws controlled around 40 percent of America's illicit supply of methamphetamines (known on the street as "crank").

Bikers gun their machines down Casino Drive in Laughlin, during the city's 20th Annual River Run.

The Super Snitch

In 1985 the FBI made a major breakthrough in its war on the outlaws when it enlisted Anthony Tait—a member of the Hell's Angels' chapter in Anchorage, Alaska—as a paid informant. Rising through the ranks, Tait became a senior officer in the Angels' national organization and supplied his FBI handlers with the low-down on drug operations and turf wars. Tait fed the FBI information for three years, until it finally swooped nationwide, arresting dozens of Angels on charges ranging from conspiracy to the possession of illegal firearms and the manufacture and distribution of illicit drugs.

Tait's evidence secured the conviction of 30 Angels for drug dealing and conspiracy. Among those imprisoned was Sonny Barger, who served five years for conspiracy to commit murder after Tait implicated him in a plot to blow up the Chicago clubhouse of rival bikers, the Outlaws. Serving his sentence in Arizona, Barger fell in love with the climate and scenery and, after his release, settled in the state's sunny Cave Creek. Now an Angels' elder statesman, he still bench presses and rides his Harley FXRT every day.

Wanted Outlaws

Federal authorities also turned the heat on the Outlaws. After being indicted on a series of racketeering charges, the club's international president, Harry "Taco" Bowman, went on the run in 1997. After two years on the FBI's "Most Wanted" list, he was arrested in 1999 and was subsequently convicted on multiple racketeering charges and sentenced to life in prison. Like the Angels, however, the Outlaws insist that, while individual members may commit crimes, the club itself is law abiding. As their website puts it—"We may not live by the rules of society, but we do live by its laws."

During his reign, Taco Bowman's operations extended across a large slice of the southern United States, bringing the Outlaws into conflict with numerous rival clubs. Enmity had long existed between the Outlaws and the Hell's Angels, but Bowman also waged a bitter war against the Hell's Henchmen, the Warlocks, and the Pagans. The Pagans, meanwhile, had their own beef with the Angels.

Hellraising Pagans

During the late 1990s outlaw clubs slugged it out for control of the Northeast seaboard, traditionally Pagan territory. But in 1999 a power vacuum had been left in Long Island when the cops busted 30 Pagans for weapons possession and conspiracy. With their rivals out of the picture, the Angels moved in.

But the Pagans didn't take things lying down. In February 2002 10 vanloads of Pagans attacked the Angels as they partied at their annual Hellraiser Ball in Plainview, New York. Wielding knives, axe handles, and baseball bats, the Pagans burst into the catering hall. The Angels responded, armed with anything ready to hand—bare fists, knives, and guns. When the battle subsided, one Pagan lay dying from gunshot wounds, while another 11 shot or stabbed bikers were rushed to the hospital.

Afterward, the police hauled in 73 Pagans on weapons and assault charges. Murder charges against a Hell's Angel were later dropped following a plea of self-defense. But the struggle for the East Coast continued as the Hells Angels pushed into traditional Pagan turf in northern New Jersey and Philadelphia, and in January 2005 Thomas Wood, vice president of the Angels' Philly chapter, was gunned down in cold blood—reputedly by a Pagan.

The Laughlin Shootout

A long and violent history of rivalry also exists between the Hell's Angels and the Mongols. Both lay claim to California, and 2002 saw several belligerent encounters between the two clubs—in San Diego and Modesto, and in Reno, Nevada.

In April 2002 the bad blood hit boiling point in the gambling town of Laughlin, Nevada. That year more than 60,000 bikers attended the 20th Annual River Run, one of the biggest motorcycle rallies in the West. The Mongols and Hell's Angels both turned up in force. After some minor skirmishes between the two clubs, the Angels decided to settle things. Just after 2.00 a.m. a column of Angels rolled up at Harrah's Casino, where the Mongols had pitched camp. There was a brief altercation between the two groups, and then a pitched battle exploded in full view of surveillance cameras. Crowds of gamblers ducked for cover as bullets whizzed by and bikers tore into each other with knives, hammers, and wrenches.

With guns drawn, dozens of lawmen stormed the wrecked casino. As they pinned cuffed outlaws to the ground, the body count became clear. Two Hell's Angels had been shot dead and a Mongol had been fatally stabbed. Sixteen other casualties were hospitalized—nine with stab or gunshot wounds and one with a fractured skull.

The "Battle of Laughlin" was a major bout in the outlaw turf wars that raged from coast-to-coast in the US throughout the 1980s and 1990s. But the clashes have not always been confined to the US. With the expansion of the outlaw clubs overseas, biker wars have stretched worldwide. Across Europe, Canada, and Australia, the biker blood feuds have taken a heavy toll.

Britain's Biker Bloodshed

A horde of outlaw biker clubs sprang up in Britain during the late 1960s—the Road Rats, the South London Nomads, the Glasgow Blues, and many others. Hell's Angels groups also appeared—but had no formal backing from the California-based club.

It was only in 1969 that official Hell's Angels charters were granted to two London chapters, followed by the addition of a West Coast chapter in 1974. But a struggle developed as the officially sanctioned Hell's Angels began pressuring the renegade clubs to join the official fold or give up their patches and disband.

The Windsor War

The errant clubs maintained their defiance and war soon erupted between the Hell's Angels factions. Hostilities with another club, the Road Rats, however, brought a truce between the two Angels groups, and in July 1976 most independent clubs agreed to become prospect chapters for the official Angels. Some, however, resisted.

The Windsor chapter was especially defiant. And, on Easter Sunday 1979, a campsite in the normally tranquil New Forest became a battlefield as the Windsor outlaws and seven official Angels chapters fought it out with knives, pistols, and shotguns—the Windsor president, Dick Sharman, only narrowly surviving after being shot through the head and chest during the mêlée.

In March 1980 prison sentences of between six months and 15 years were passed on 29 Hell's Angels for their part in the battle. But sporadic violence continued until 1983, when a coup in the Windsor camp deposed Sharman, and the Windsor club finally joined Hell's Angels, England, as a prospect chapter.

Above and top right: British bikers of the late 1960s often emulated the style and attitude of the American outlaws.

Two Sides of the English Outlaws

Growth during the 1980s and 1990s saw the Hell's Angels become Britain's leading outlaw motorcycle club, with 16 chapters based throughout England and Wales. Compared with their brothers overseas, the English Angels have a relatively law-abiding profile, winning friends through their annual Bulldog Bash—one of Europe's biggest motorcycle and music festivals. In 2002 the English Angels even got a royal seal of approval when a club member rode in an official procession celebrating the Queen's Golden Jubilee.

But there is still an ugly side to the English outlaws, evidenced by a violent feud during the late 1990s. The expansion of the Outcasts—a biker outfit based in London and Great Yarmouth (on the east coast of England)—put them at loggerheads with the Hell's Angels. The struggle culminated in 1998 in southwest London at Battersea's Rocker's Reunion. Usually a good-humored festival, the Reunion turned nasty as a small group of Outcasts were ambushed by a 40-strong squad of bikers. Two Outcasts were killed as the attackers waded in with knives, machetes, axes, and hammers.

Showdowns in Australia

Club rivalries and struggles for territory have given Australia its own share of brutal biker wars. Between 1972 and 1975 a bitter feud between the Hell's Angels and the Rebels brought a spree of vicious confrontations. But Australia's worst single incident of "bikie" violence came during the 1980s, in what became known as the "Milperra Father's Day Massacre."

Massacre at Milperra

When a Sydney chapter of the Comancheros defected to the Bandidos in 1984, the Comanchero president, William "Jock" Ross, was livid. The feud simmered for several months before detonating in a fierce shootout in Milperra, a quiet suburb of Sydney.

At Milperra's only pub—the Viking Tavern—bikers had organized a party to celebrate Australia's Father's Day on September 2nd. When word got to Jock Ross that the Bandidos would be there, his Comancheros armed up and went to settle the score. For a few minutes there was a standoff, but then the powder keg exploded in a bloody battle featuring guns, knives, machetes, and baseball bats. When the smoke cleared, the bar's parking lot was littered with the dead and dying. Six bikers and a 14-year-old girl had been killed, while more than 20 people lay seriously wounded.

The court case that followed was one of the largest in Australian history—with seven counts of murder brought against 43 people. At the trial's conclusion Jock Ross and seven other Comancheros received life sentences for murder, while 16 Bandidos were sentenced to 14 years each for manslaughter.

Operation Panzer

The Milperra Massacre pushed Australia's bikie clubs into the spotlight and the authorities began getting tough. Concerned that Aussie outlaws were developing into sophisticated criminal organizations, the police took concerted action against the clubs. In 1995 Australia's National Crime Authority launched Operation Panzer, an investigation of biker involvement in organized crime; and by 2001 the initiative had resulted in more than 900 charges. Nevertheless, the Australian police argue that outlaw clubs such as the Rebels, the Gypsy Jokers, the Hell's Angels, the Black Uhlans, and the Finks remain a criminal force to be reckoned with, responsible for the manufacture and national distribution of illicit drugs worth millions of Australian dollars.

In Australia, as in the United States, bikers wear their colors with pride.

The Great Nordic Biker War

An especially violent and sustained biker war rocked Scandinavia during the mid-1990s. What police sources later dubbed "The Great Nordic Biker War" began with brief outbursts of violence, but ultimately resulted in 11 murders, 74 attempted murders, and 96 people wounded as battle raged between 1994 and 1997.

The hostilities were focused on Denmark. The country had been a Hell's Angels' stronghold since the club's first Danish chapter was established in 1980. But in 1993 things began to change when the Undertakers, a Danish outlaw club affiliated to the Bandidos, looked to expand. With the Bandidos set to extend their influence across Scandinavia, vicious rivalry developed between the newcomers and the Hell's Angels. Some observers have suggested the turf war was simply about status and reputation. But others have argued it was a fierce struggle for control of Copenhagen's criminal drugs network.

Outlaws on the Warpath

The campaign kicked off in 1994 in the Swedish port of Helsingborg, where three Hell's Angels were wounded and one of their supporters killed in a gun battle with Bandidos. Sporadic shootouts and bombings followed across Finland, Norway, Sweden, and Denmark. When antitank missiles (stolen from a Swedish army depot) were fired at Hell's Angels' clubhouses in Helsingborg (Sweden) and Helsinki (Finland), it looked as though the Bandidos were getting the upper hand. But the Hell's Angels struck back. In March 1996 Bandido leader Uffe Larson was shot and killed at Copenhagen airport—six Angels were later convicted of his murder.

The Bandidos responded in kind. In April 1996 more antitank rockets were fired at Hell's Angels' clubhouses in Sweden and Denmark, and the ensuing months saw gun, bomb, and grenade attacks on members of either side. In September 1996 there was a series of machinegun and rocket attacks on Hell's Angels' clubhouses. Then, in October, a missile scored a direct hit on the Hell's Angels' Copenhagen headquarters, killing two and wounding 16. Further sporadic shootouts and tit-for-tat hits followed, and in June 1997 a car bomb totally destroyed the Bandido's Drammen clubhouse in Norway, wounding 22 and killing a woman nearby.

Making the Peace

The war was taking a heavy toll and government ministers across Scandinavia began to legislate against motorcycle outlaw activity. Then, either because they were sick of the cycle of violence or because they wanted to avoid a massive police crackdown, the two clubs agreed to a ceasefire brokered by Thorkild Høyer (a famous Danish lawyer). In September 1997 the Bandidos' European president, "Big" Jim Tinndahn, and the Hell's Angels' leader Bent "Blondie" Nielsen appeared on Danish TV news—the two biker chieftans shaking hands to confirm an end to the four-year feud.

An Uneasy Truce

In September 2003, however, a former Bandido and convicted kidnapper, Mickey Borgfjord Larsen, was killed by a car bomb in a Copenhagen suburb. For a moment it looked as if the infamous biker war had resumed, but fears subsided as suspicion quickly fell on Larsen's own erstwhile associates. Generally, the situation

In the Scandinavian biker war of the mid-1990s, outlaw funerals became commonplace.

in Scandinavia stabilized after the 1997 "peace treaty," although relations between the Hell's Angels and the Bandidos have remained fragile ever since.

The truce didn't totally convince the Danish authorities. Tight legislation curbing biker activity remains in force, and police surveillance of the clubs continues amid strong suspicions that the two gangs have carved up control of Copenhagen's drug trafficking, arms dealing, and prostitution rackets between themselves. The "Great Nordic Biker War" is over, but the peace is uneasy.

The Oirsbeek Nomads Murder Case

Other European police forces have also suspected the increasing involvement of outlaw bikers in organized crime. A particularly curious case unfolded in The Netherlands in 2004. In February of that year the bullet-riddled bodies of three Hell's Angels of the Oirsbeek Nomads chapter (including their president, Paul de Vries) were found dumped in a stream near the city of Echt. Rumors quickly spread that the triple murder was related to a failed drugs deal. Furthermore the plot took a strange twist when police raided the Nomads' clubhouse and rescued two bikers from Curacao (an island in the Dutch Antilles renowned for cocaine smuggling) who were allegedly being held hostage.

Details of the case remain unclear and the links between the murders and the alleged kidnapping are still hazy. But in August 2004 the press reported that the 12 remaining Nomads club members had been charged with being part of a criminal organization—the first time an entire Hell's Angels' chapter, rather than individual members, has been officially linked to organized crime in the Netherlands.

"Mom" Boucher and the Battle for Montreal

The international expansion of America's outlaw motorcycle clubs had its most devastating consequences in Canada. In 1994 a brutal biker war hit the streets of Montreal as the Hell's Angels wrestled with rival outlaws the Rock Machine for control of Quebec's billion-dollar drug trade. The war took a devastating toll, the next six years seeing no less than 162 killings, 200 attempted murders, and scores of wounded.

The Brotherhoods Square-Up

The Hell's Angels' first Canadian chapter was established in 1977, in Montreal, during the club's first wave of international expansion. By 1985, two more chapters had been added in Quebec and, according to police investigators, had taken control of a large slice of Montreal's drug trade. The operation was ruled with a rod of iron and—in what became known as the "Lennoxville Massacre"—five Angels suspected of squandering their brothers' profits were summarily executed.

During the late 1980s the Montreal Hell's Angels coexisted alongside another biker club, the Rock Machine, established by Salvatore Cazzetta in 1986. In 1994, however, Cazzetta was arrested for attempting to import 11 tons of cocaine. For the Angels, it seemed like a good opportunity to make a move.

Led by Maurice "Mom" Boucher, the Montreal Angels began to muscle in on the Rock Machine's operations. The hits began in July 1994 when a Rock Machine associate was gunned down in a downtown motorcycle store. Another was blown in half by a car bomb that also killed an 11-year-old boy playing nearby. The following month the first Hell's Angel was shot and killed at a shopping mall, and on the day of his funeral nine bombs exploded around the province. The next months saw the war intensify as the Hell's Angels and the Rock Machine fought for control.

The Hell's Angels vs. The Rock Machine

A skillful tactician, Mom Boucher mustered his forces, forming a new Nomads chapter from the toughest Hell's Angels in the region. The Rock Machine, in contrast, was about half the size of the Angels, with around 60 members and associates in its two chapters (Montreal and Quebec City). With Cazzetta still in prison, the ambitious Fred Faucher took command and made overtures to the Bandidos, hoping to secure backing from a major firm. But in 1997, when the Bandidos' international vice president traveled to Canada to check out the Rock Machine, the police pounced. The Bandido leader was deported, while Faucher and 22 Rock Machine members were arrested (though the Rock Machine eventually secured Bandidos' backing in 2000).

The Angels were also feeling the heat. In 1995 the police formed the "Wolverines," a crack unit of 60 detectives dedicated to breaking the biker gangs. Protest marches thousands strong were organized to demand tough action from the authorities, and the Canadian government passed laws beefing up the cops' powers to combat organized crime.

Open Warfare

Boucher seemed to take the crackdown as a personal challenge. In June and September of 1997, two off-duty prison guards were murdered in drive-by shootings. Hitman Stephane Gagné was later arrested and testified that Boucher had ordered the killings to intimidate his opponents. Boucher was arrested, but in November 1998 was acquitted of the murders—a jury evidently refusing to believe Gagné's evidence. Walking free, Boucher looked untouchable and flaunted his apparent impunity around Montreal's nightspots.

The violence also became more brazen. In September 2000, veteran crime journalist Michel Auger published an exposé of biker-related crime on the front page of his newspaper, *Le Journal de Montreal*. The next day he was shot six times and only narrowly survived the assassination attempt.

Montreal Hell's Angels leader Maurice "Mom" Boucher seemed untouchable after he was acquitted of the murder of two prison guards in 1998.

Toward the end of 2000 things turned sour for Mom Boucher. In October he was rearrested for the prison-guard shootings. The government had won the right to retry him and he was soon charged with 13 more killings and other crimes. Another blow for the Montreal Angels came in March 2001. Following the largest investigation in Canadian history, 120 bikers and their associates were arrested in a series of police raids known as "Operation Springtime."

Boucher was found guilty of murder in May 2002 and received a life sentence. For the other club members, a series of "mega trials" was conducted. In September 2003 nine Quebec Hell's Angels received prison terms ranging from 10 to 15 years for drug-trafficking and conspiracy to commit murder, and in February 2004 warrants were issued for the arrest of a further 63 Angels.

"THE LAST CHAPTER"

The Canadian bikers' turf war was spectacularly brutal and bloody, but it also added to the media's fascination with the outlaw biker. In 2002—just as the Montreal Angels' reign of terror was ending—a $10-million biker mini-series had its debut on Canadian TV. The "Last Chapter" shifted the action from Montreal to Toronto, but the plot was eerily familiar. It featured Michael Ironside as an imperious outlaw boss who oversees the expansion of a criminal biker empire.

RoadKing
UltraClassic

08

ROLLING THUNDER

The Iron Horse Corral

The Great American Freedom Machine

Harley-Davidson JD, 1926

Harley's first 1200cc engine was introduced in 1922, dubbed the "Superpowered Twin." The JD model had a full generator system and lights. The 1926 version of the JD had a three-speed transmission and a 3.5-gallon (16-liter) fuel tank. A complete tool kit was also standard equipment.

Engine Side-valve V-Twin
Capacity 1200cc
Top speed 85+ mph (136+ km/h)

Bikes, sickles, hogs—whatever you want to call them, motorcycles are the lean, mean incarnation of individuality and freedom. Nothing can match the sensory rush of speed and power as you gun your bike down an open road. While most passions fade with time, the thrill of a speeding motorcycle never wavers. Big, bad, and beautiful, the motorcycle is the ultimate embodiment of rebellious cool.

Harley-Davidson EL, 1939

Launched in 1936, the EL was Harley's first big V-Twin with overhead valves and a re-circulating oil system. The appearance of the new engine was especially striking. The prominent rocker boxes—looking like knuckles on a fist—soon earned the engine the nickname "Knucklehead."

Engine Overhead-valve 45° V-Twin
Capacity 1000cc
Top speed N/A

Old-School Hogs

More than any other ride, it's the Harley-Davidson—the hefty, heroic "hawg"—that reaches out to a biker's visceral impulses. Other bikes might be faster or more advanced technologically, but nothing can touch the magnificent looks and sound of the Harley.

The old-school Harley design dates back to the 1936 "Knucklehead." Named for the bulbous rocker boxes on top of the engine, the "Knucklehead" was Harley's first overhead valve big "V" twin. Refinements ensued, and in 1941 a bigger 1200cc version (the FL) was added to the Milwaukee stable.

In 1948 the "Panhead" was born, named for its panlike valve covers. Hydraulically damped, telescopic forks were also introduced in 1949 with the 74FL Hydra-Glide. Launched in 1958, the Duo-Glide saw the addition of rear hydraulic suspension, while the 1965 Electra Glide introduced Harley's first electric starter. In 1957 the XL Sportster was launched to do battle against the racy foreign competition, and laid the way for the illustrious line of Sportster models that followed. In 1966, meanwhile, improvements to the Panhead engine spawned the "Shovelhead"—named for its rocker-box covers that resembled the heads of coal shovels.

Harley-Davidson F, 1946

The "Knucklehead" engine was Harley's workhorse throughout the 1930s and 1940s. Steadily refined, it was the solid, reliable muscle behind such bikes as Harley's 1946 "F" model.

Engine Overhead-valve 45° V-Twin
Capacity 1200cc
Top speed N/A

"The Best of Times" and "The Worst of Times"–The AMF Years

Harley-Davidson U, 1938

In 1937 Harley introduced the "U" model, with a four speed transmission and a recirculating oil system as standard. With its frame, brakes, clutch, and transmission all beefed up in 1938, the 74 U model became a popular choice for sidecars and commercial use.

Engine Side-valve V-Twin
Capacity 12000cc
Top speed N/A

Some classic Harley models were produced during the 1960s, but facing competition from foreign imports, Harley needed cash to update its antiquated production line.The solution came in the form of a corporate buyout, with the American Machine and Foundry Company (AMF) officially taking control of Harley-Davidson in January 1969 for around $21.6 million. Under AMF's aegis, Harley-Davidson's production quickly tripled, but quantity was at the expense of quality, and the AMF Harleys became notorious for their dismal production standards. However, the AMF management sanctioned many respected developments in the Harley lineup.

Harley-Davidson FL, 1947

The FL series first appeared in 1941 with the launch of a 1200cc "Knucklehead" model. In 1946 a style makeover breathed new life into the FL, as Art Deco curves of the 1930s gave way to more clear-cut lines.

Engine Overhead-valve 45° V-Twin
Capacity 1200cc
Top speed N/A

Willie G.'s Custom Classics

At the design helm, William G. Davidson hung out with Harley-Davidson's following. His affinity with street-level riders helped him pioneer a new generation of "factory customs"—Harleys that emulated the look and feel of a mean, muscular outlaw machine. Introduced in 1971, the FX Super Glide was the first in the pedigree, pairing the Electra Glide's 1200cc engine and frame with the Sportster's chopper-inspired front end and a Euro-style "boat tail" fender. In 1977 there followed the sleek FXS Lowrider, a meaty cruiser that featured rakish front forks, drag-style handlebars, alloy wheels, and a cool-looking, low-slung saddle. The FXEF Fat Bob of 1979 also looked pretty mean, with its chunky dual gas tanks and bobbed fenders. In 1980 the trend continued with the FXWG Wide Glide and the FXB Sturgis—the latter being a limited edition machine (based on the Low Rider) released to commemorate the annual biker blowout held at Sturgis. Harley also flirted with café-racer design, launching a short-lived sportbike—the XLCR—in 1977. The XLCR looked the part, but couldn't match the performance of European and Japanese sportbikes, so Harley's street-legal roadracer lasted just two years.

Harley-Davidson FXB Sturgis, 1980

The "Sturgis" name first appeared on a Harley-Davidson in 1980 as a part of the "FX" Super Glide line. The FXB Sturgis was a limited edition bike released to commemorate the annual Sturgis bike rally. Based off the FXS Low Rider, the FXB had a belt primary and final drive—hence the "B" designation. The belt drive proved its worth and Harley soon added it to several other models.

Engine Overhead-valve 45° V-Twin
Capacity 1340cc
Top speed 105 mph (170 km/h)

The Eagle Flies Alone—Harley-Davidson Reborn

June 1st, 1981 is a date venerated by Harley-Davidson aficionados. For it was on this day that the company finally severed its ties with AMF. The 11-year relationship had never been a happy affair and in 1981 13 Harley executives (including Willie G.) finally raised around $80 million to buy out the company. Proclaiming "The Eagle Flies Alone," a newly energized Harley-Davidson took full advantage of its defiant aura and relaunched itself as a bold, all-American trailblazer.

But times were still hard. During the 1970s the Japanese had scooped a large slice of the American big-bike market, and Harley's share slipped from 80 percent to less than 30 percent. By 1985 the firm was sliding into debt and only narrowly escaped bankruptcy. The late 1980s, however, saw the beginning of a major revival. Temporary assistance had come in the form of a 45 percent tariff on Japanese motorcycles larger than 750cc, while in 1983 Harley began rebuilding customer relations with the launch of HOG—the Harley Owners Group—a company-sponsored "club" that drew Harley riders into the corporate fold.

Exciting new models also boosted Harley-Davidson's fortunes. In 1984 the new 1340cc Evolution engine—or "Evo"—retained the classic V-twin look, but its updated technology allowed for more power and smoother running. With the Evolution came a whole new generation of machines. Especially successful was the FXST Softail, which had suspension springs hidden under the frame, giving it the appearance of a chopped 1950s hardtail. Indeed, a slick "retro" look became a Harley trademark. Unveiled in 1986, for example, the FLST Heritage Softail capitalized on the classic style of the 1950s big twin Hydra-Glide. Even more impressive was 1990's FLSTF Fat Boy. Designed by Willie G. himself, the Fat Boy won a firm following with its "nostalgic" styling combined with a powerful Evolution engine.

Into the New Millennium

Staying a jump ahead of the competition, Harley introduced the new Twin Cam 88 engine during the mid-1990s, a powerhouse that retained the classic air-cooled twin arrangement, but with a fresh twin cam design and 8 extra cubic inches (from 80 to 88—hence the designation). A bigger break with tradition came with the introduction of the V-Rod in 2001. Based on Harley's VR-1000 racing bike, the V-Rod was Harley-Davidson's first motorcycle to boast a 60-degree splay of cylinders, water-cooled motor, and wraparound frame. Opinion on the V-Rod, however, was divided. Some loved its power, handling, and flamboyant curves, but the die-hards were uncertain about the radical styling.

There was, however, no doubting Harley-Davidson's return to success. The firm enjoyed almost 10 years of increasing sales until 2003, its 100th anniversary. The Motor Co. celebrated its centenary in style, with a year-long series of weekend festivals roaring into 10 cities worldwide and a three-day party in Milwaukee that saw a million Harley fanatics flood into the city.

In 2003 Harley-Davidson celebrated its 100th anniversary and, at the beginning of a new millennium, the authentic Harley Hog remains the biker's solid, steadfast steed.

STURGIS

The Noble Indian

Indian Chief, 1920s

Introduced in 1922, the Chief was originally intended to downsize the big bike, offering 1200cc performance from a high-tech 1000cc engine. But, in response to customer demand, 1923 saw Indian introduce the larger 1200cc "Big Chief."

Engine Side-valve 42° V-Twin
Capacity 1000cc
Top speed 90 mph (144 km/h)

Indian motorcycles was the only American motorcycle manufacturer to compete seriously against the might of Harley-Davidson. Founded in 1901, Indian went from strength to strength, culminating in 1913/1914, when the company produced nearly 32,000 motorcycles a year and accounted for a 42 percent share of the entire American motorcycle market.

Indian Chief, 1950s

Steadily refined, the Chief was in production into the 1950s, when probably some of Indian's best bikes rolled off the production line. But the company finally bit the dust in 1953.

Engine Side-valve 42° V-Twin
Capacity 1300cc
Top speed 120 mph (192 km/h)

The Classic Tribe

One of the company's biggest successes was the Scout. Beginning life in 1920 as a 600cc, mid-sized machine, the Scout later appeared as a 750cc version and as a 101 Scout with an improved frame, suspension and steering. But Indian's big machine flagship was the Chief. Launched in 1922, the Chief had a large—62 cubic inch—V-Twin engine, upgraded to a 74-inch "Big Chief" version in 1923. Steadily refined over the years, the Chief remained in production into the 1950s, with the launch of an 80 cubic inch Blackhawk Chief in 1951.

Back on the Warpath

In the late 1990s the Indian name was revived. A limited edition Indian Chief was produced in 1999, followed by a Centennial Chief in 2001 to celebrate Indian's 100th anniversary. With the company firmly re-established, more gleaming Indians rolled off the production-line in the new millennium.

Indian Chief, 2000

With the relaunch of the Indian marque in the late 1990s, the legendary Chief returned to the road. Limited production began in 1999, with successive new models following in the new millennium.

Since 2000 the Chief has been available in a variety of colors, with enough options to match any rider's sense of individuality and personal expression.

Engine Overhead-valve V-Twin
Capacity 1400cc
Top speed N/A

The Bulldog Breed: British Bike Legends

BSA Gold Star 1958

The "Gold Star" marque was introduced by BSA to commemorate the Gold Star awarded at the Brooklands racetrack to any rider who achieved a lap of over 100 mph (160 km/h). In 1937 Wal Handley lapped at 107.57 mph (173 km/h) on a standard BSA Empire Star 500, and his success inspired the launch of BSA's first "Goldie" in 1938. Throughout the 1950s and early 1960s successive versions of the Gold Star were an ever-popular choice among Britain's coffee-bar cowboys.

Engine Overhead-valve Single cylinder
Capacity 500cc
Top speed 110 mph (177 km/h)

US bikers have tended to favor all-American thoroughbreds, but British bikes have always won respect. During the 1950s and 1960s a large proportion of Britain's motorcycle production was geared to export sales—especially for the US market, where the lighter, faster Norton, BSA, and Triumph machines gave the American workhorses a run for their money.

1956 Norton Dominator 99

Designed by Bert Hopwood and released in 1949, the Dominator was Norton's first parallel twin. In 1952 the engine was slotted into a featherbed frame and released as the Dominator 88, while in 1955 the 600cc Dominator 99 was added to Norton's stable. In 1962 the Dominator was upgraded to the 650SS model—a racing beast with enough muscle to challenge Triumph's sports models.

Engine Parallel Twin
Capacity 600cc
Top speed 110 mph (177 km/h)

The Glory Days

BSA enjoyed huge success during the 1950s and the famous Goldstar became the stuff of legend, scooping a bevy of race trophies. The 1960s saw café racers bolt from the BSA stable—most notably the 1965 Spitfire, a 650cc sports twin (built until 1968).

The 1950s and 1960s also saw success for Norton. The Dominator, Norton's first parallel twin, was launched in 1949, the engine slotted into the famous Featherbed frame and released as the Dominator 88 in 1952. But the most famous Norton of all was the parallel-twin 750cc Commando, launched in 1968. Popular in the United States, a bigger 828cc version was introduced in 1973.

But Triumph was Britain's biggest success story. Triumph's 500cc TR5 Trophy was a race-winner throughout the late 1940s and 50s, while the 650cc Thunderbird became a biker icon when Marlon Brando appeared astride one in *The Wild One*. The 650cc Bonneville and 500cc Tiger Daytona were also the stuff of legend.

The Bulldog Bites Back

Despite their solid reputation and racetrack victories, British motorcycle firms were on the rocks by the late 1960s. But Triumph began to claw its way back in the 1990s. New Daytona, Trophy, Trident and Thunderbird models were a hit in the showroom, their sleek styling evoking the firm's heyday of the 1950s and 1960s.

Triumph Thunderbird sport

Triumph's 650cc Thunderbird 6T was originally launched in 1949. Designed by Edward Turner, the maestro of British motorcycle design, the Thunderbird was named after a "Thunderbird Motel" in South Carolina, where Turner once stayed. After years in the corporate wilderness, the 1990s saw Triumph back in business. The Thunderbird, once again, was its flagship model following the launch of a new "T" Bird in 1994. In 1998 a Thunderbird Sports model was added, with stunning looks that harked back to the classic café-racer style of the 1960s.

Engine In-line Triple cylinder
Capacity 885cc
Top speed 137 mph (220 km/h)

Banzai!: The Japanese Invasion

The Japanese motorcycle invasion had humble origins. In 1948 Soichio Honda had hit on the idea of fitting small, army-surplus engines into bicycle frames, with a belt drive powering the rear wheel. Meeting Japan's voracious demand for cheap transportation, Honda's concept was a money-spinner.

In 1949 Honda moved to a factory in Hamamatsu, switching production to four-stroke, 150cc bikes. Other lightweight machines followed, most notably 1958's C100, with its characteristic step-through frame. In all its assorted variants, the "Honda 50" would go on to become the world's best-selling motorcycle, with sales of around 30 million bikes. But it was Honda's break into the American market that sealed the company's success.

"You Meet The Nicest People on a Honda"

Honda opened an American office in 1959 and set to work selling its brand of zippy little bikes to young Americans looking for a fun runabout. By December 1962 Honda was selling more than 40,000 motorcycles a year in the US, but bigger things were to come. During the early 1960s the company's "You Meet The Nicest People On a Honda" ad campaign promoted the Honda 50 as the perfect pair of wheels for everyone—housewives, young couples, and (especially) teenagers. It was a masterstroke. Whereas Harleys were fast becoming associated with greasy outlaw trash, Honda presented its bikes as a font of bubbly fun, and by 1970 the company was selling over 500,000 machines a year in the US alone.

Honda's Asian confederates followed suit. Yamaha, originally founded in 1887 as a musical-instrument manufacturer, diversified into motorcycle manufacture in 1955. Its first bike, the YA1, was a 125cc two-stroke and was soon followed by 175cc and 250cc machines. Kawasaki adopted the same strategy. Kawasaki Heavy Industries had started life as an industrial conglomerate that built ships, aircraft, and trains. After World War Two the company focused its output on consumer demand, building a series of small, two-stroke motorcycles. Likewise, Suzuki (set up in 1909 by Michio Suzuki as a textile company) shifted its attention to the consumer market during the 1950s, producing its first motorcycle in 1952 with the launch of its 36cc Power Free motorized bicycle.

By the early 1960s Yamaha, Kawasaki, and Suzuki had all joined Honda in a US export drive—all four companies targeting hip and happening American youngsters. The promotion of motorcycles as fun and friendly invigorated the market and pushed up sales of the Japanese imports. Sales of Harley hogs, meanwhile, were diminishing daily. But Harley's fortunes took an even steeper dive when the Japanese began zeroing in on the big-bike market.

Big Bike Samurai

In 1964 Honda began edging into midsized bike production with the launch of the 300cc CB77, followed by the bigger CB450 in 1965—a machine whose racing version was ridden to victory at Daytona in 1967. But it was Honda's CB750 that really took the industry by surprise. No one had really believed the Japanese were capable of competing in the big-bike market, but the CB750 blew its slower, less reliable, and more expensive rivals off the map.

The Japanese also made inroads into heavyweight touring territory. Launched in 1974, Honda's Gold Wing was a massive Winnebago of a motorcycle—the engine of the 1991 version displacing a huge 1520cc. One thing missing from its list of luxury options, however, was attitude.

But the Japanese manufacturers also took a shot at the outlaw image. Honda, Kawasaki, Yamaha, and Suzuki have all offered faux Harleys and long, low "factory customs." Yet somehow, for most US bikers, the passion has never been there—for all their reliability, they still lack the character and soul of an honest-to-goodness Hog.

With emphasis on fun and excitement, Honda made a play for the American motorcycle market during the 1960s with a host of smaller, relatively cheap bikes—including the Scrambler 450.

The Custom Bike Renaissance

The 1990s and 2000s have seen a phenomenal revival of the American motorcycle industry. In 2004 retail sales of bikes in the US were up 7 percent. Obviously, Harley-Davidson was the major beneficiary, the Milwaukee Motor Co.'s profits growing 17 percent as more than 317,000 Hogs rolled off the production line.

A New Breed of V-Twins

But the growing enthusiasm for bikes also kickstarted a new generation of smaller motorcycle firms who offered their own take on the all-American ride. Equally impressive are the machines produced by the American IronHorse Co. For example, founded in Fort Worth in 1995 by Bill Rucker and Tim Edmondson, IronHorse produce an awesome range of seven custom V-twins, as well as a Signature Series line for "once-only" orders. In Colorado, meanwhile, the Panzer Motorcycle Works have been "putting the past into the present" since 1996 with their range of superb retro-styled choppers—including magnificent reproductions of the bikes from *Easy Rider.*

JESSE JAMES AND WEST COAST CHOPPERS

Since founding West Coast Choppers in 1993 in the corner of his mom's garage, Jesse James has shot to celebrity status. From handcrafting exhaust pipes, fenders, and gas tanks that look more like stylish sculptures than motorcycle parts, his company has grown into a 50-person, 18,000-square-foot (1,675-square-meter) workshop with an estimated annual turnover of $6 million.

Each one of Jesse James's handmade bikes is custom tailored to its owner. Though he also has a few trademark touches—including incredibly stretched front forks and 140-bhp engines machined from a solid block of aluminum. Not to mention the .44 magnum and 9 mm shell casings adorning the gas-cap covers and riser bars—a nod to Jesse's outlaw namesake. Mounting up on a West Coast Chopper, however, comes at a price. One of Jesse's bikes is likely to set you back as much as $150,000. Yet there's no shortage of customers—the waiting list is usually a year long.

American Eagle are in the vanguard of the custom motorcycle renaissance. Loaded with chrome accessories, Streetfighter is a classy Bad Boy.

Bill Hayes, current National Press and Publicity Officer of Boozefighters MC and author of the book, *The Original Wild Ones*. The bike is a replica of the Captain America chopper from *Easy Rider*, with an engine designed and built by the world-class motorcycle builder, Berry Wardlaw.

The Custom-Ride Master Builders

The new Golden Age of American motorcycling has also brought a new generation of custom builders. Ron Simms, Arlen Ness, and Roger Bourget are still at the leading edge of the custom-bike scene, but they have been joined by a legion of new master builders. Most notably Jesse James, owner of West Coast Choppers, and the Teutuls (Paul Snr. and Paul Jr.) at Orange County Choppers, have won a devoted following of fans who love their badass biker image and the drop-dead cool of their custom-built motorcycles.

American Eagle also exemplify a new legion of firms specializing in limited runs of "custom production motorcycles"—bikes with the look and performance of a custom ride, but assembled by professional manufacturers. Founded in 1995, American Eagle has become a leader in the field of custom, high-performance cruisers.

At over 9 feet (2.7 m) long, and featuring a 45-degree rake, American Eagle's Raptor has that classic "old school" chopper feel.

ORANGE COUNTY CHOPPERS

Established in 1999 Orange County Choppers began as an offshoot of Orange County Ironworks, a steel-fabrication business owned by custom-chopper builder Paul Teutul Snr. Based in Rock Tavern, New York, Teutul made his mark at the Daytona Biketoberfest in 1999 with "True Blue"—a classic chopper built in his basement. Other custom machines followed, including the "Jet Bike," the "Spider Bike," the "Black Widow," and the "Fire Bike" (built to honor the New York City firefighters who lost their lives at Ground Zero).

In 2002 Orange County Choppers got national attention with the launch of the Discovery Channel TV series, *American Chopper*. Each episode of the "fly-on-the-wall" documentary captures the company's day-to-day dramas as Paul Snr. and his sons battle tight deadlines (and invariably one another) to create some of America's most stunning custom motorcycles.

09

AN ENDURING HERO

The Biker Rides Again

The Legend Lives On

Since the late 1960s—when the maverick motorcyclist was a byword for violence and outrage—the biker has been steadily rehabilitated. Some of the outlaws may still be linked to crime and bloodshed, but big bikes and badass style are now a multimillion dollar business. And in popular culture the image of the biker has been reborn—the feral outsider recast as a cherished personification of honest, authentic Americana.

Motorcycle Retro

The biker's makeover began during the mid-1970s, amid a wave of nostalgia for the 1950s. Movies such as *American Graffiti* (1973) and *Grease* (1978) rediscovered the romance and fun of hot rods, drive-ins, and rock 'n' roll—and the biker, too, returned to the screen as a hero who embodied sexy rebellion. In *The Lords of Flatbush* (1974) Sylvester Stallone and Henry Winkler (then obscure unknowns) starred as young, leather-clad roughnecks feeling their way toward adulthood against the backdrop of 1957 Brooklyn. But it was Winkler's role in TV's *Happy Days* that was the biker hit of the 1970s. The actor played Arthur Fonzarelli, or simply "The Fonz," a motorcycle-riding high priest of cool.

During the 1980s the image of the 1950s biker maintained its enigmatic appeal in retro movies such as *The Loveless* (1982) and *Rumble Fish* (1983). Meanwhile, the strong, silent heroes of sci-fi action movies such as the *Mad Max* trilogy (1979, 1981, 1985) and the *Terminator* series (1984, 1991, 2002) confirmed the enduring magnetism of the tough, leather-jacketed loner.

"FONZIE"

As the super-cool 1950s biker, Arthur "Fonzie" Fonzarelli, actor Henry Winkler starred in *Happy Days*, ABC TV's longest-running sitcom. After the success of the 1973 movie, *American Graffiti*, ABC programers were looking to cash in on the surge of 1950s nostalgia and premiered *Happy Days* in January 1974. Initially, ABC and Paramount (the program's makers) thought the show's appeal would be short-lived—but the series proved a winner, and over the next 10 years it ran for 11 series in 255 episodes.

Originally *Happy Days* centered on the escapades of two fun-loving, high-school kids—Richie Cunningham and his pal Warren "Potsie" Weber—as they grew up in 1950s Milwaukee. Richie (played by Ron Howard, who had starred in *American Graffiti*) was supposed to be the innocent teenager while Potsie (Anson Williams) was his more worldly pal. Fonzie, a hip biker dropout, was originally on the sidelines, but the character was a hit with viewers and the show increasingly focused on the relationship between the ultracool Fonzie and the hopelessly straight Richie and his friends.

Over the years "The Fonz" rode a selection of Harleys and Triumphs and his thumbs-up gesture, gift for effortless cool—"Heeeeyyyh!"—and black biker jacket became *Happy Days* trademarks. Initially, however, TV executives were nervous of the delinquent overtones of Fonzie's leather jacket and motorcycle boots, and the show's first few episodes saw Fonzie wearing a windbreaker and penny loafers. Gradually, the network agreed Fonzie could wear a leather jacket when he was riding his motorcycle and by the end of season one it had become a firm fixture. In 1980 one of Fonzie's leather jackets even became an exhibit in the Smithsonian Museum of American History in Washington, DC.

"Heeeeyyyh!"—in TV's *Happy Days*, the leather-jacketed Fonzie was the embodiment of biker cool.

The Rumble of the Rolex Riders

America's renewed romance with the biker has spawned a new breed of motorcyclist. A posse of well-heeled and middle-aged rebels—accountants, lawyers, doctors—has hit the road astride gleaming Harleys, affecting a Bad Boy look with classy leathers and designer boots. More at home in the boardroom than the saloon, they cut loose at the weekends, gunning their $40,000 custom-made machines down the highway.

The motorcycle passions of millionaire publisher Malcolm Forbes were indicative of the bike's developing upmarket appeal. Before his death in 1990, Forbes' biker associations were an important part of his larger-than-life reputation. Aged in his 40s, Forbes had taken up motorcycling during the late 1960s, riding with a club called (with more than a hint of irony) the Capitalist Tools, and purchasing a cycle dealership in New Jersey that developed into one of America's largest. Forbes' international riding trips also made him a goodwill ambassador for motorcycling, and in 1987 he was rewarded with the American Motorcycle Association's Hazel Kolb Brighter Image Award—the AMA's highest accolade for those who generate good publicity for the motorcycle world.

The upmarket Rolex Riders or RUBs ("Rich Urban Bikers") are often reviled as role-playing impostors by the blue-collar, outlaw die-hards. But, while there is no mistaking the motorcycle's new mainstream appeal, the biker himself will never be entirely tamed. Beneath the gloss there remains a fierce edge to the biker's image, a kick of exciting rebellion that will always set the iron horse faithful apart from the predictable, conventional majority.

Billionaire Malcolm Forbes (left) typified a new affluent, middle-aged breed of "Rolex Rider." In 1987 he presented actress Elizabeth Taylor with "Purple Passion"—a custom-crafted Harley Sportster.

The Rebel Flame Keeps Burning

In the twenty-first century, biker culture remains alive and well. Harley-Davidson's centenary year was 2003, and to commemorate the occasion wave upon wave of polished chrome thundered into Milwaukee, the Harley heartland. In the new millennium, sales of motorcycles are surging ahead, and every year hundreds of thousands of bikers clad in black leather and wraparound shades converge at uproarious rallies in Sturgis, Laconia, Daytona, Myrtle Beach, and countless other cities and towns around the world to talk bikes and live the biker lifestyle. Solid, defiant, and supremely cool, the biker is set to remain an all-American icon of resilient individuality and freedom.

The last American hero—the biker walks tall as an icon of individualism and robust independence.

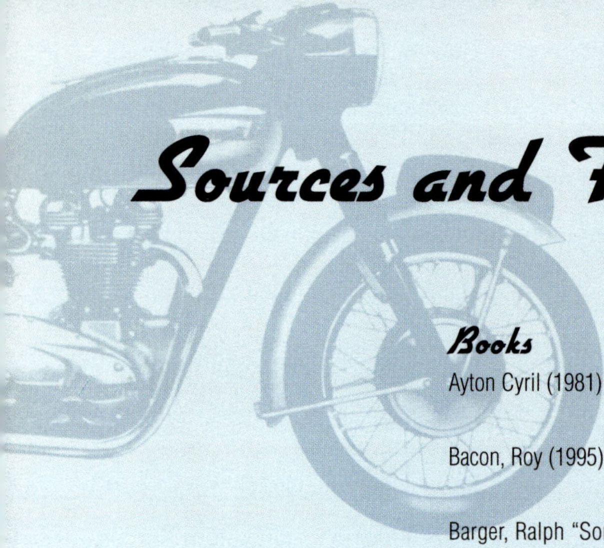

Sources and Further Reading

Books

Ayton Cyril (1981) *Japanese Motorcycles*, London: F. Muller.

Bacon, Roy (1995) *British Motorcycles of the 1960s*, London: Osprey.

Barger, Ralph "Sonny" (with Zimmerman, Keith and Zimmerman, Kent) (2001) *Hell's Angel: The Life and Times of Sonny Barger and the Hell's Angels Motorcycle Club*, London: Fourth Estate.

Barnes, Richard (1979) *Mods!*, London: Eel Pie.

Charles, Gary (2003) *Bikers: Legend, Legacy and Life*, London: Independent Music Press.

Clay, Mike (1988) *Café Racers: Rockers, Rock 'n' Roll and the Coffee-Bar Cult*, London: Osprey.

Dicks, Shirley (2002) *Road Angels: Women Who Ride Motorcycles*, New York: Writers' Club.

Farren, Mick (1985) *The Black Leather Jacket*, London: Plexus.

Ferrar, Ann (2001) *Hear Me Roar: Women, Motorcycles, and the Rapture of the Road*, North Conway: Whirlaway.

Ganneau, Didier and Dumas, Francois-Marie (2001) *A Century of Japanese Motorcycles*, Osceola, WI: Motorbooks International.

Garson, Paul and the Editors of Easyriders (2003) *Born To Be Wild: A History of the American Biker and Bikes, 1947–2002*, New York: Simon & Schuster.

Harris, Maz (1985) *Biker: Birth of a Modern Day Outlaw*, London: Faber and Faber.

Hayes, Bill (2005) *The Original Wild Ones: Tales of the Boozefighters*, Osceola, WI: Motorbooks International.

Holmstrom, Darwin and Nelson, Brian J. (2002) *BMW Motorcycles*, Osceola, WI: Motorbooks International.

Joans, Barbara (2001) *Bike Lust: Harleys, Woman, and American Society*, Madison: University of Wisconsin Press.

Lavigne, Yves (1990) *Hells Angels: 'Three Can Keep A Secret If Two Are Dead'*, New York: Carol Publishing.

Lavigne, Yves (1990) *Hells Angels at War*, Toronto: HarperCollins.

Lavigne, Yves (1996) *Hells Angels: Into the Abyss*, New York: Harper.

Levingstone, Tobie Gene (with Zimmerman, Keith and Zimmerman, Kent) (2003) *Soul On Bikes: The East Bay Dragons MC and the Black Biker Set*, St. Paul, MN: MBI.

Lyon, Danny (2003) *The Bikeriders*, San Francisco: Chronicle.

Mayson, Barry and Marco, Tony (1985) *Fallen Angel: Hell's Angel to Heaven's Saint*, New York: Doubleday.

Paradis, Peter (2002) *Nasty Business: One Biker Gang's Bloody War Against the Hells Angels*, Toronto: HarperCollins.

Rafferty, Tod (2001) *The Indian: The History of a Classic American Motorcycle*, London: Salamander.

Reynolds, Frank and McClure, Michael (1969) *The True Story of Hell's Angels by a Hell's Angel*, London: New English Library.

Reynolds, Tom (2000) *Wild Ride: How Outlaw Motorcycle Myth Conquered America*, New York: TV Books.

Sato, Ikuya (1991) *Kamikaze Biker: Parody and Anomy in Affluent Japan*, Chicago: University of Chicago Press.

Seate, Mike (2000) *Two Wheels on Two Reels: A History of Biker Movies*, North Conway: Whitehorse Press.

Seate, Mike (2003) *Choppers*, St. Paul, MN: MBI.

Seate, Mike (2004) *Outlaw Choppers*, St. Paul, MN: MBI

Shaylor, Andrew (2004) *Hells Angels Motorcycle Club*, London: Merrell.

Sher, Julian and Marsden, William (2003) *The Road to Hell: How The Biker Gangs are Conquering Canada*, Toronto: Alfred Knopf.

Sucher, Harry (1981) *Harley-Davidson: The Milwaukee Marvel*, Newbury Park: Haynes Publications.

Stuart, Johnny (1987) *Rockers!*, London: Plexus.

Tanaka, Rin (2000) *Motorcycle Jackets: A Century of Leather Design*, Atglen, PA: Schiffer.

Thompson, Hunter (1966) *Hell's Angels: The Strange and Terrible Saga of the Outlaw Motorcycle Gangs*, New York: Random House.

Upright, Michael (1999) *One Percent*, Glendale, CA: Action Publishing.

Vanderheuvel, Cornelius (1997) *A Pictorial History of Japanese Motorcycles*, Osceola, WI: Motorbooks International.

Veno, Arthur (2002) *The Brotherhoods: Inside the Outlaw Motorcycle Clubs*, Crows Nest NSW: Sue Hines.

Walker, Mick (1994) *Café Racers of the 1960s*, London: Windrow & Greene.

Wethern, George and Colnett, Vincent (1978) *A Wayward Angel*, New York: Richard Marek.

Wilde, Sam (1977) *Barbarians on Wheels*, London: New English Library.

Wolf, Daniel (1992) *The Rebels: A Brotherhood of Outlaw Bikers*, Toronto: University of Toronto Press.

Wright, David (2003) *The Harley-Davidson Motorcycle: A 100-Year History*, North Branch, MN: CarTech.

Yates, Brock (1999) O*utlaw Machine: Harley-Davidson and the Search for the American Soul*, Boston: Little Brown and Company.

Sources and Further Reading

Magazines

Back Street Heroes—www.backstreetheroes.com/

Bikers News—www.bikersnews.de/index_02.html

Easyriders—www.easyriders.com/

The Horse—www.ironcross.net/

Motorpsycho!—www.ratbike.org/motorcycho/mchopage.html

Outlaw Biker—www.outlawbiker.com/

Ozbike—www.ozbike.com.au/

Street Chopper—www.streetchopperweb.com/

100% Biker—www.100-biker.co.uk/about.htm

Websites

American Motorcycle Association—www.ama-cycle.org/

Bandidos MC—www.bandidosmc.com/

BikerGazette.Com—bikergazette.com/

Bikernews.Net—www.bikernews.net/index.cfm

Bikers Magazine—bikersmag.com/

Boozefighters MC—www.bfmcnatl.com/

David Mann Art Gallery—home.att.net/~knucklehead-47/dmann.htm

Harley-Davidson—www.harley-davidson.com/selector.asp

Hells Angels MC—www.hells-angels.com/

Indian Motorcycles—www.indianmotorcycle.com/

The Original Wild Ones—www.theoriginalwildones.com/

Outlaws MC—www.outlawsmc.com/index1.htm

Pagans MC—hometown.aol.com/jailedpagans/

Museums

American Motorcycle Museum
Zwolsestraat, 63C 8101 AB, Raalte, Netherlands
www.ammh.nl/

Barbour Vintage Motorsports Museum
2721 5th Ave. S., Birmingham, AL. 35210
www.barbermuseum.org/

BMW Museum
Petuelring 130, Munich, Germany
www.bavaria.com/culture/bmwmuseum_ge.html

Canadian Vintage Motorcycle Museum
housed within the Canadian Military Heritage Museum
347 Greenwich St., Building #19,
Brantford, Ontario, Canada
www.cvmg.on.ca/museum.php

Fukuyama Auto and Clock Museum
(75 motorbikes in collection) 3-1-22, Kita-yoshizu-cho,
Fukuyama City, Hiroshima
720-0073 JAPAN
www.facm.net/eng/index2_e.html

Harley-Davidson Museum
1425 Eden Rd., York, PA 17402
www.h-dmuseum.com/

Indian Motorcycle Museum
33 Hendee St., Springfield, MA 01109

London Motorcycle Museum
Oldfield Lane South, Greenford,
Middlesex UB6 9LD, UK
www.motorcycle-uk.com/lmm.htm

The Motorcycle Hall of Fame Museum
13515 Yarmouth Dr., Pickerington, Ohio 43147
www.ama-cycle.org/museum/index.asp

Museo Nazionale Del Motociclo
Via Casalecchio, 58/N,
47900 Rimini, RN, Italy
www.museomotociclo.it/

National Motorcycle Museum
Coventry Road, Bickenhill, Solihull,
West Midlands,B92 0EJ, UK
www.nationalmotorcyclemuseum.co.uk/

The National Motorcycle Museum
33 Clarkson Street,
Nabiac, NSW, Australia
www.nationalmotorcyclemuseum.com.au/info.html

The Pioneer Museum of Motorcycles
P.O. Box 2351,
Tacoma, WA 98401
www.museumofmotorcycles.com/default.htm

Sturgis Motorcycle Museum and Hall of Fame
P.O. Box 602, Sturgis,
South Dakota 57785
www.sturgismotorcyclemuseum.org/

Sveriges Motorcykelmuseum
Gyllenhjulet Surahammar, Sweden
www.autosite.se/mchkmalardalen/Eng%20index.htm

Trev Deeley Motorcycle Museum
13500 Verdun Pl.,
Richmond, BC V6V 1V2, Canada
www.trevdeeley.com/home.html

Wheels Through Time Museum
1121B Veteran's Memorial Dr.
Mt. Vernon, Il 62864
www.wheelsthroughtime.com/

Index

Index

Acknowledgments

A big thanks to all the usual family and friends. Thanks, too, to Mark Fletcher for sorting things out, and to all the crew at Ivy Press—Sophie Collins, Jason Hook, and Caroline Earle. Thanks also go to Simon Goggin, Shelley Noronha, and Karl Shanahan for helping bring everything together. And a special thanks to the Bandidos MC and Bill Hayes, Boozefighters MC for all their help.

Picture Credits

The publishers wish to thank Paisano Publications, LLC, and in particular Mark Dodge, General Counsel and Director of Licensing, John Nielsen, photo editor, and Kim Peterson, editor of *In the Wind* magazine and photo archivist for their extraordinary assistance in providing content from the archives of *Easyriders* magazine, and other information helpful to the Ivy Press in researching this work. The publishers would also like to thank the following organizations and individuals for their kind permission to reproduce the photographs in this book. Every effort has been made to acknowledge the pictures, however we apologize if there are any unintentional omissions.

The Advertising Archive: 13, 158, 159.
Allied Artists Corporation: 120 top.
Alvey Towers (www.alveyandtowers.com): 156–157 all images.
Arizil Realty and Publishing Company, Arizona (*SEE* magazine): 71.
Bike Photo Library (www.bikephoto.co.uk): 154–155 all images.
Steve Blackwell, Triton Owners' Club President: 121 bottom.
BMC-USA: 132.
BOOZEFIGHTERS/Bill Hayes, author of *The Original Wild Ones* and National Press & Publicity Officer for the Boozefighters MC: 8, 26–29 all images, 160–161 top.
British Film Institute: 48–49, 51 top left, 63 top right, 66 right, 78 left, 101, 120 center right and bottom.
Choppers Magazine, Maywood: 81.
Columbia Pictures Corporation: 32.
CORBIS: 6 (© Neal Preston), 15 (Minnesota Historical Society), 16 bottom (Underwood & Underwood), 33 top right (Bettmann/Corbis) 148–151 all images (Tria Giovan), 102 (Sunset Boulevard/Corbis Sygma), 107 (Bettmann/CORBIS), 111 (© Patrick Ward), 141 (Nik Wheeler).
EMPICS: 126 (AP), 136 (DPA), 152 top left (POLFOTO).
Em Tee Publications New York (*Men Today*): 73.
The Gazette, Montreal: 145.
Getty Images: 9 (Time Life Pictures), 14 both images (Hulton Archive), 16 top left (Hulton Archive), 17 (Hulton Archive), 18–25 all images (Hulton Archive), 34, 35 (Time Life Pictures/Frank Scherschel), 36 (Hulton Archive), 44 (Time Life Pictures/Bill Ray), 45 top left, 51 top right (Hulton Archive), 60 (Hulton Archive), 62 left, 62–63 center (Hulton Archive), 65 top right (Hulton Archive), 80 (Time Life Pictures/Allan Grant), 82 (Time Life Pictures/Ralph Crane), 87 top right (Paul Harris), 91 (Hulton Archive), 106 (Hulton Archive), 110 (Hulton Archive), 114 (Hulton Archive), 116 left (Hulton Archive), 117 (Time Life Pictures/Terrence Spencer), 118 left (Hulton Archive), 118–119 (Hulton Archive), 121 top (Hulton Archive), 127 (Hulton Archive), 133 (Time Life Pictures), 146 (Scott Olsen), 152–153 center (Scott Olsen), 153 (Scott Olsen),162 (Scott Olsen), 166 (Time Life Pictures), 167 (Scott Olsen).
Hellkats MC Midwest: 113.
Jay and Ell Enterprises Inc. Ohio: 88.
The Kobal Collection/Columbia: 33 top center, 64, 65 top center.
Las Vegas Sun/Ethan Miller: 138.
Chris Ledford, Webmaster, Bad Toys Holdings, Inc., Parent company of American Eagle Motorcycles, Inc.: 160 bottom left, 161 bottom right.
Motorcycle Hall of Fame Museum Photo Collection: 105.
National Screen Service Corp.: 104.
NBC News: 130–131.
New Zealand Herald: 100.
Robert Opie: 12 top.
Paisano Publications, LLC: 46, 81 top right, 92, 93, 94, 95, 112 left.
Popperfoto.com: 10, 33 top left, 96–97 center.
QMG Magazine Corp, New York (*Man's Conquest*): 72.
Rex Features /EVT: 45 top right, 54 left, 98.
Rolling Stone: 96 left (Michael Maggia, © 1970 *Rolling Stone*).
San Francisco Chronicle/Barney Peterson: 30–31.
Saturday Evening Post/Wayne Miller: 47.
Spirit of *Oz* (www.richardneville.com): 86, 87 center.
Sports Illustrated: 79 top (Heinz Klutmeier).
Topham Picturepoint: 84–85 center (Topfoto/Imageworks), 134 (Topfoto), 137 (Topfoto), 140 both images (Topfoto), 142 (Topfoto/Polfoto), 143 (Topfoto/Polfoto).
David Winogrond: 61 top right.